THE SOLAR SYSTEM IN MINUTES

GILES SPARROW

Quercus

The solar system

11

7

9

1 Sun
Rotation period: 25–35 days
Diameter: 1,391,700 km
(864,400 miles)

2 Mercury
Orbital period: 88 days
Rotation period: 58.6 days
Diameter: 4,878 km (3,030 miles)

3 Venus
Orbital period: 224 days
Rotation period: 243 days
Diameter: 12,104 km (7,518 miles)

4 Earth
Orbital period: 365.25 days
Rotation period: 24 hours
Diameter: 12,756 km (7,923 miles)

5 Mars
Orbital period: 687 days
Rotation period: 24.6 hours
Diameter: 6,787 km (4,216 miles)

6 Ceres (dwarf planet within the main asteroid belt)
Orbital period: 4.6 Earth years
Rotation period: 9.1 hours
Diameter: 975 km (606 miles)

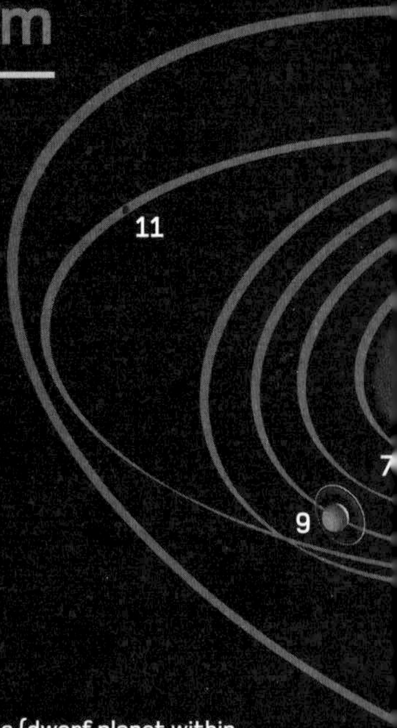

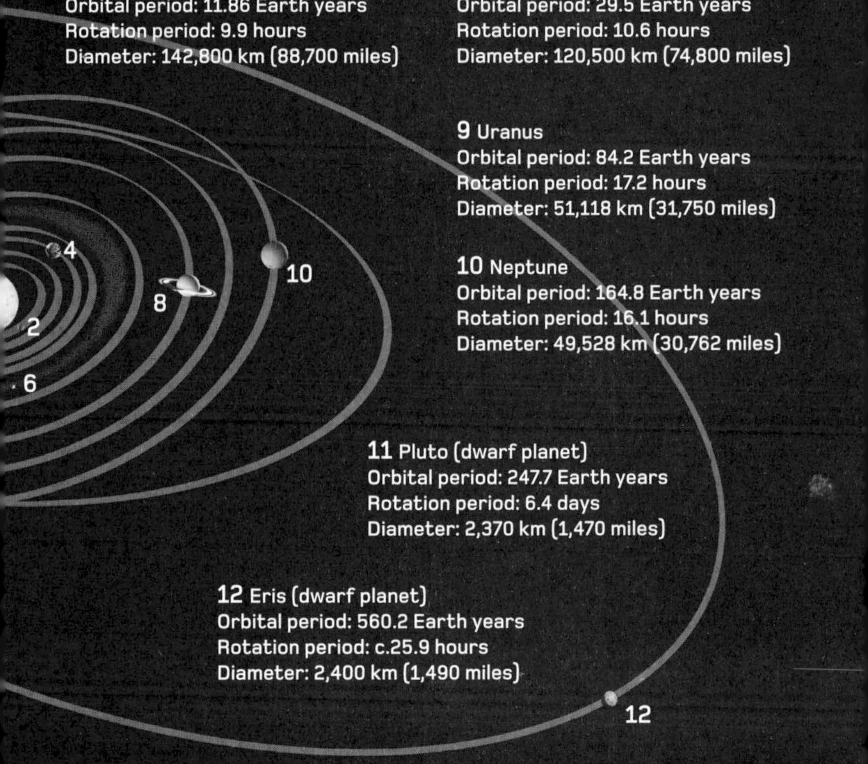

7 Jupiter
Orbital period: 11.86 Earth years
Rotation period: 9.9 hours
Diameter: 142,800 km (88,700 miles)

8 Saturn
Orbital period: 29.5 Earth years
Rotation period: 10.6 hours
Diameter: 120,500 km (74,800 miles)

9 Uranus
Orbital period: 84.2 Earth years
Rotation period: 17.2 hours
Diameter: 51,118 km (31,750 miles)

10 Neptune
Orbital period: 164.8 Earth years
Rotation period: 16.1 hours
Diameter: 49,528 km (30,762 miles)

11 Pluto (dwarf planet)
Orbital period: 247.7 Earth years
Rotation period: 6.4 days
Diameter: 2,370 km (1,470 miles)

12 Eris (dwarf planet)
Orbital period: 560.2 Earth years
Rotation period: c.25.9 hours
Diameter: 2,400 km (1,490 miles)

4

10

8

2

6

12

CONTENTS

Introduction

Our solar system is the region of the Universe about which we know the most – an array of planets, moons and smaller bodies held in orbit around a huge ball of exploding gas that provides them with heat and light. It's the one region of the cosmos that we can visit directly – currently using robotic probes, but in the future through human exploration as well.

The more we learn, the more complex the solar system becomes – each of its planets and major moons has a unique evolutionary history, preserved in geological scars or internal chemistry. Even smaller objects such as asteroids and comets show surprising variety as we learn more about them. It is also now clear that the solar system is a dynamic place – its objects might be separated by vast gulfs of empty space, but over timescales of millions of years or more, close encounters and even collisions are fairly common. Computer simulations are now revealing how the orbits of the planets and their moons have evolved over time, and how similar processes continue to influence asteroids and comets.

The Solar System in Minutes is a concise yet comprehensive guide to the wonders of our cosmic back yard, from bloated giant planets that swallow comets whole, to volcano-tortured moons and deep-frozen outcasts in 10,000-year orbits. Every major object (and a significant number of minor ones of special interest) has an entry – if not a chapter – to itself, broadly in order from the Sun outwards. Planet Earth, famously described as the 'third rock from the Sun', of course has its own chapter, but many other worlds we shall visit also have much to tell us about our home. Some preserve episodes in the solar system's shared history; some reveal secret mechanisms that are obscured on Earth's ever-changing surface; and others serve as 'what ifs?' – cautionary tales of how our planet might have turned out if things had been just slightly different.

A final chapter offers a brief review of our explorations in the solar system so far, and prospects for the future. Crewed missions to other worlds will be expensive and dangerous, but are surely inevitable. They will certainly make scientific discoveries we cannot yet imagine, and by stretching the limits of our technological capability may prove our species worthy of long-term survival – not as residents of one small, vulnerable ball of rock, but as citizens of the wider Milky Way galaxy.

What is the solar system?

Most astronomers would agree with a broad definition of the solar system as the region of space dominated by the influence of the Sun. Beyond that, however, differences soon emerge over exactly how to define the Sun's 'dominance'. Some take a more limited view that the solar system extends as far as the heliopause, a region where the stream of particles driven out from the Sun

on the solar wind comes to a halt in the face of pressure from an 'interstellar medium' made of stellar winds from countless other stars. By this definition, solar influence only reaches about 123 times further than Earth's orbit around the Sun, and NASA's Voyager spaceprobes have therefore already left the solar system.

A more inclusive definition, however, is based on the dominance of the Sun's gravity. This extends the solar system almost a thousand times further, out to the most remote comets that orbit the Sun at distances of about one light year. This is the definition used for objects that merit inclusion in this book.

Scale of the solar system

The distances between planets are huge when considered in everyday terms – if you could get into a car and simply drive to the Sun at an average speed of 100 km/h (60 mph), the journey would take 170 years. For this reason, astronomers use much larger units of measurement when considering distance in the solar system. The astronomical unit (AU) is defined as the average distance between the Earth and the Sun, 149.6 million km (93 million miles). In terms of AU, the orbits of the planets range from innermost Mercury (an average of 0.39 AU from the Sun, though this varies considerably – see page 70), out to Neptune (an average of 30.11 AU from the Sun). The asteroid belt (which divides the solar system into rocky inner planets and giant outer planets) encircles the Sun between 2 and 3.3 AU, while the Kuiper Belt of icy dwarf planets is thought to extend to about 100 AU. The distant Oort Cloud forms a shell around the solar system out to a distance of about 63,000 AU, or 1 'light year' – its outer edges are so far away that light from the Sun takes a year to reach them.

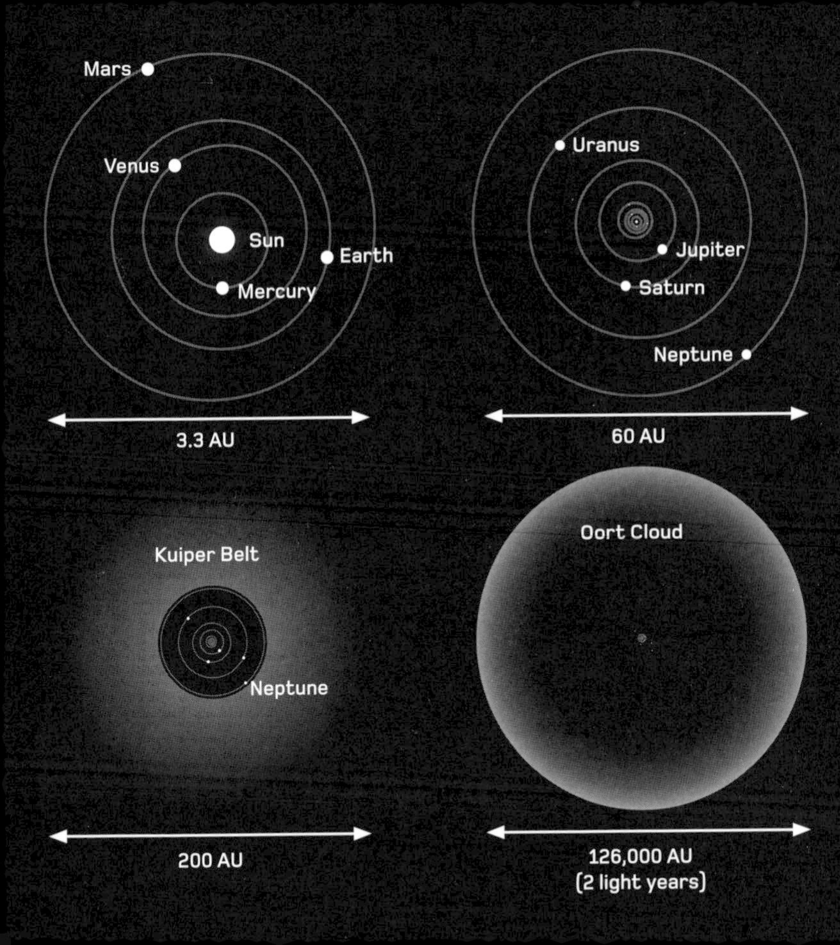
Mars
Venus
Sun
Earth
Mercury
3.3 AU
Uranus
Jupiter
Saturn
Neptune
60 AU
Kuiper Belt
Neptune
200 AU
Oort Cloud
126,000 AU
(2 light years)

Orbits and gravity

The Sun's gravity is the dominant force in the solar system. Physically, gravity is a force exerted by objects with mass that draws other objects towards them. However, in practical terms, the effect of this attraction is often to maintain less massive objects on a more or less stable path *around* more massive ones – a path called an orbit. All stable orbits take the form of an ellipse – a circle stretched along one axis with two points called 'foci' to either side of its centre. In a 'two-body' system where one object is much more massive than the other (for example the Sun and its planets, or planets and their orbiting satellites or 'moons'), the more massive object sits at one focus and the less massive one orbits around it (in more evenly balanced systems, both objects orbit around a common 'centre of mass'). The closest approach between the two objects is known as periapsis (perihelion in the case of orbits around the Sun) and the widest separation as apoapsis (aphelion). Orbiting bodies move faster near periapsis and more slowly at apoapsis, and, as a general rule, the further out an object orbits, the more slowly it moves.

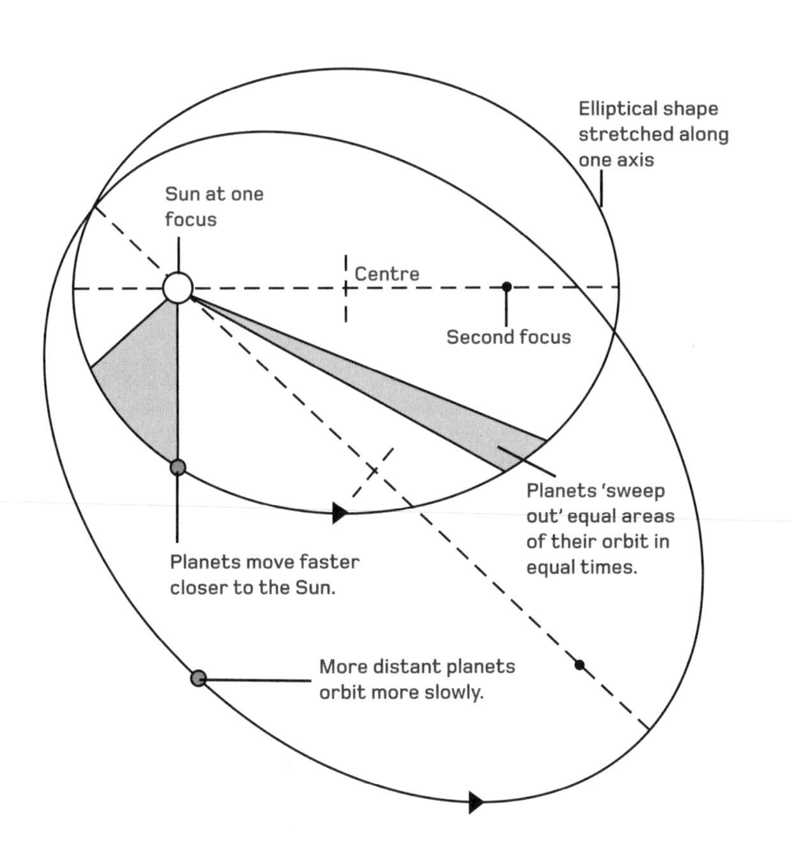
Elliptical shape stretched along one axis
Sun at one focus
Centre
Second focus
Planets 'sweep out' equal areas of their orbit in equal times.
Planets move faster closer to the Sun.
More distant planets orbit more slowly.

Our place in the Milky Way

Our Sun is just one of about 200 billion stars in an enormous spiral galaxy called the Milky Way. This disc-shaped system is about 100,000 light years across and a few hundred light years thick. It has a pronounced 10,000-light-year bulge at its centre, where stars are closely packed together, ultimately in orbit around a monster black hole with the mass of 2.6 million Suns. Although stars are more or less evenly distributed across the disc, the bright regions where new stars are born form spiral arms that wind their way out across the disc.

The solar system is located about halfway across the galactic disc, on the outer edge of a minor spiral arm called the Orion Spur. At 26,000 light years from the core, the Sun orbits the galaxy every 240 million years. From our location inside the disc, we can see a marked difference in the concentration of stars in different directions. Where we look across the galactic disc, stars line up behind each other in huge clouds. They are sparsely scattered, however, where we look 'up' or 'down' out of the disc.

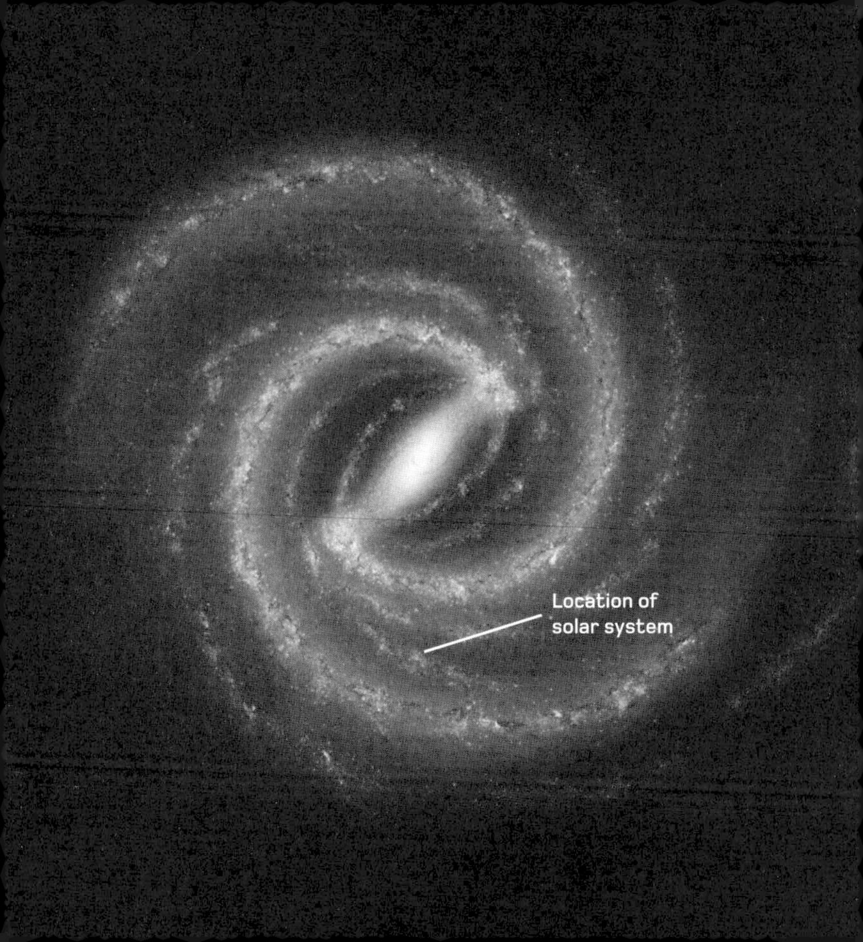

Location of
solar system

Defining planets

The word 'planet' comes from the Greek for 'wanderer', an indication of the way that certain bright objects in the night sky – our neighbouring worlds Mercury, Venus, Mars, Jupiter and Saturn – first stood out through their movements against the more distant 'fixed' stars. When Uranus, Neptune and Pluto were discovered following the invention of the telescope, they were all classed as planets. By the late 20th century, however, astronomers realized that Pluto was just one of many icy objects orbiting beyond Neptune. The discovery of a large object now called Eris (see page 362) raised the dilemma of either adding dozens of new planets in the coming decades of discovery, or agreeing to a more formal, scientific definition. Hence today, according to a ruling agreed in 2006, a planet is officially defined as a body in an independent orbit around the Sun, with sufficient mass to be spherical and enough gravitational influence to 'clear its orbit' of large interlopers. Worlds such as Pluto, Eris and the large asteroid Ceres, which meet only the first two criteria, are termed 'dwarf planets'.

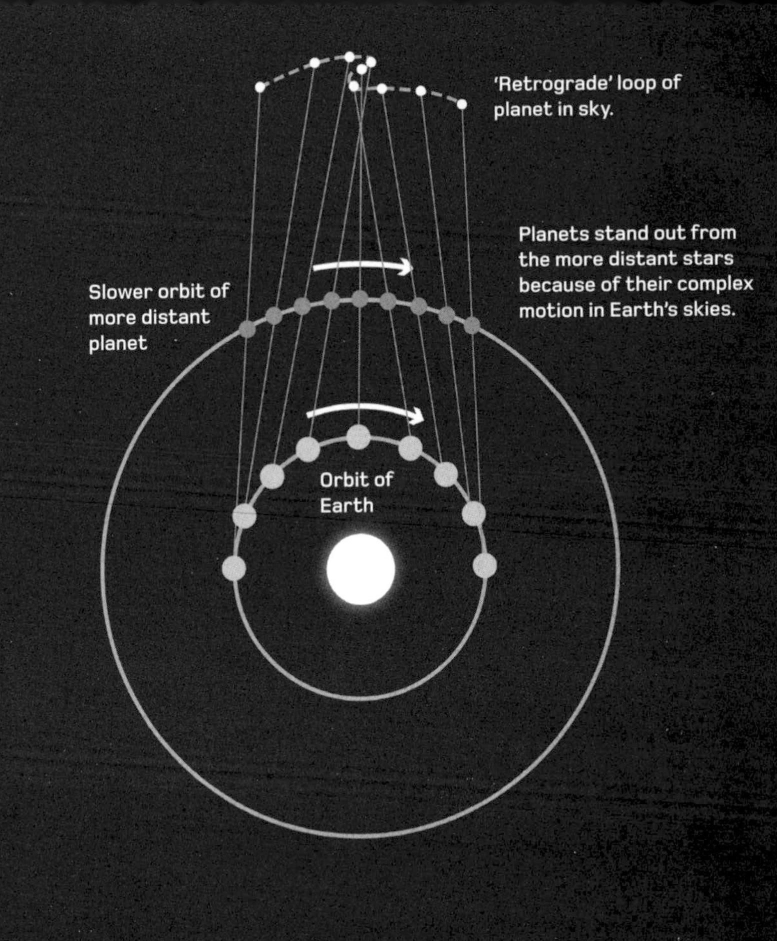
'Retrograde' loop of
planet in sky.
Planets stand out from
the more distant stars
because of their complex
motion in Earth's skies.
Slower orbit of
more distant
planet
Orbit of
Earth

Rocky planets

The planets of the solar system are broadly divided into rocky inner worlds and giant outer ones (see page 20). The rocky planets Mercury, Venus, Earth and Mars (as well as Earth's large natural satellite, the Moon) are all composed principally of materials with high melting points (atmospheric gases, surface water and ice on Earth and Mars are minor when compared to the bulk of the planets).

The collisions that formed these worlds heated most of their rocks to melting point, allowing them to separate into layers with the heaviest elements (predominantly iron and nickel) sinking to the centre and the lightest rising upwards to form a crust. In between lies an intermediate mantle region that transfers heat from the core towards the surface (either through simple conduction or through a slow process of convection in which hot rocks actively rise through their surroundings). Broadly speaking, the smaller worlds have cooled more rapidly while the larger ones retain more heat and, therefore, show more geological activity on their surfaces.

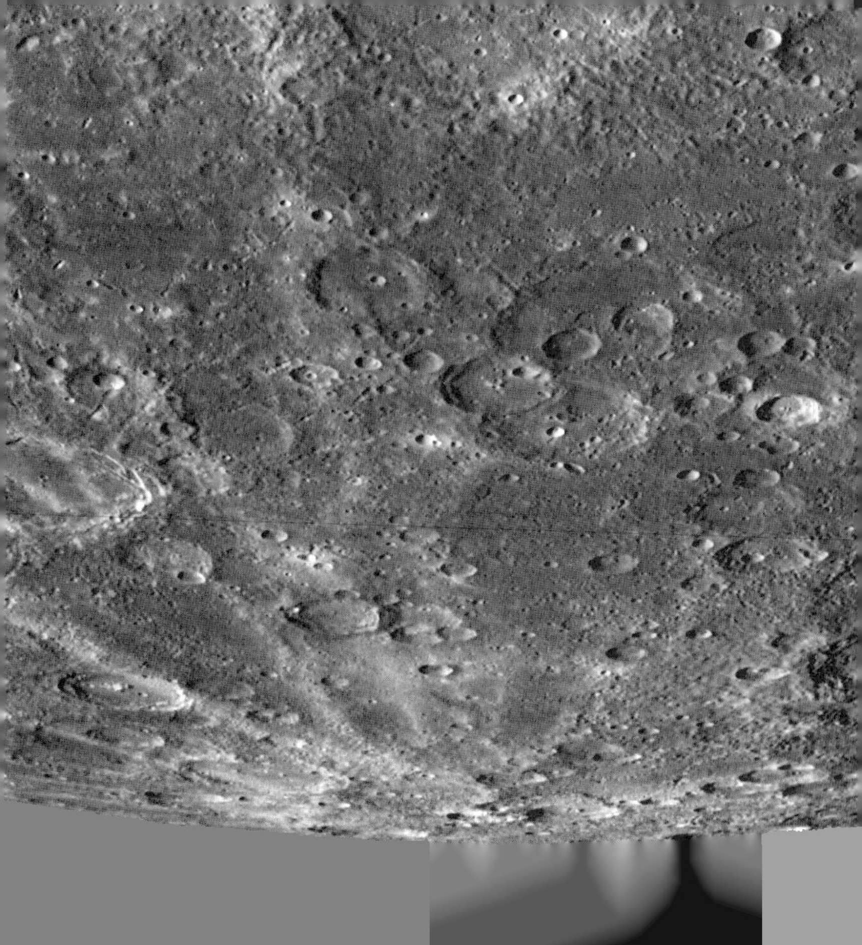

Giant planets

The planets of the outer solar system are much larger and much more widely separated than their inner neighbours. Dominated by materials with relatively low melting points (which were unable to persist closer to the heat of the Sun), they have relatively small solid cores surrounded by huge envelopes of gas, liquid or slushy materials.

The giants are divided into two pairs – closer to the Sun lie enormous Jupiter and Saturn, true 'gas giants' composed mostly of the lightweight gas hydrogen, which transforms into vast liquid seas a few thousand kilometres beneath their visible cloudtops. Further out lie Uranus and Neptune, considerably smaller (though still much larger than Earth) and rich in relatively complex chemical 'ices' – not only water (H_2O) but also ammonia (NH_3), methane (CH_4) and others. Although the term 'gas giants' is still casually used to encompass all four planets, scientists increasingly make the distinction of calling Uranus and Neptune 'ice giants'.

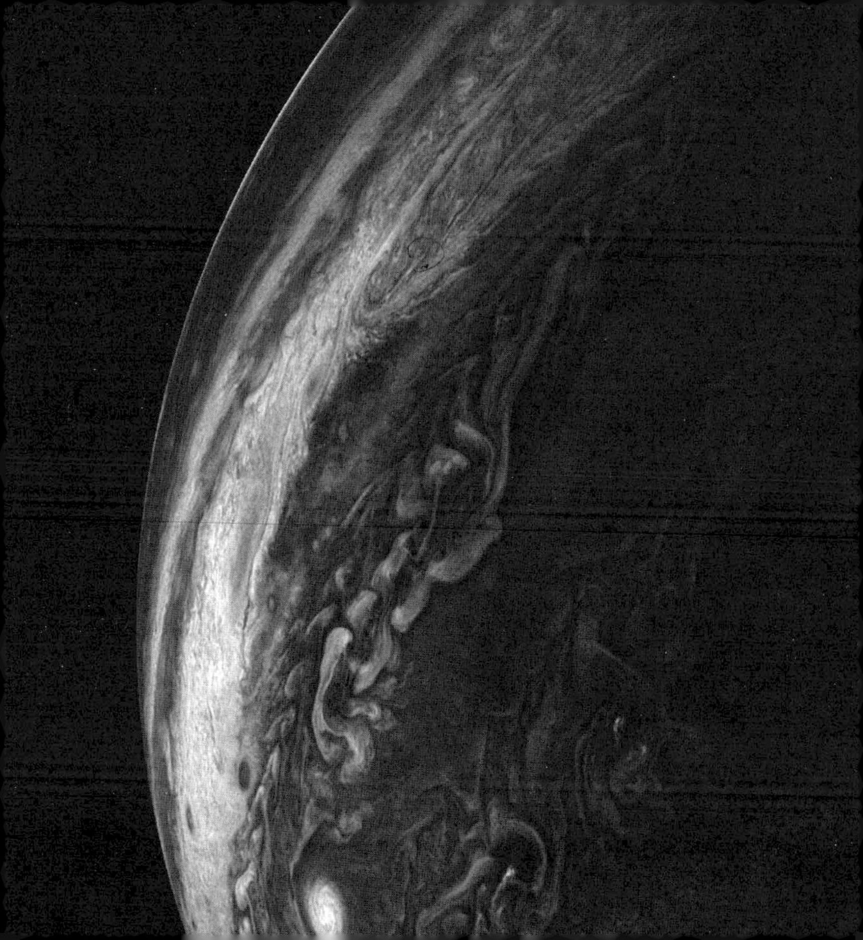

Moons and rings

Not every object in the solar system directly orbits the Sun – many instead orbit the major planets or smaller worlds. These range from substantial moons in their own individual orbits to ring systems – swarms of smaller particles that follow concentric circular orbits in a single narrow plane. Moons that follow close-to-circular orbits, near the plane of a planet's equator and in the same direction as the planet's own rotation, are known as 'regular' satellites – with a few exceptions, such as Earth's own Moon, they are thought to have condensed from material left behind as their parent planet formed. Moons in more elliptical or highly tilted orbits, or those which orbit the 'wrong way' around a planet, are called irregular satellites – they mostly started out as comets or asteroids in independent orbits around the Sun, before being captured by their planet's gravity. The particles that form ring systems, meanwhile, are generally thought to be the shattered remnants of earlier moons that fragmented due to interplanetary collision or gravitational stress.

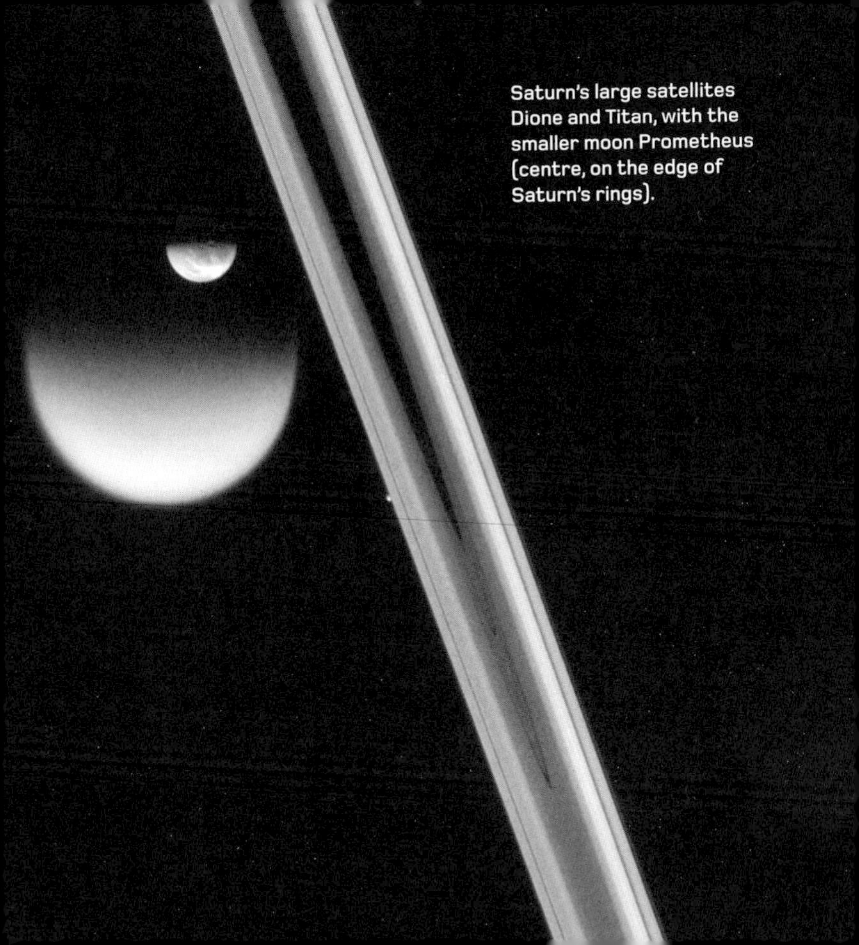

Saturn's large satellites Dione and Titan, with the smaller moon Prometheus (centre, on the edge of Saturn's rings).

Asteroids and comets

In addition to planets and the moons and rings in orbit around them, the solar system is filled with countless other objects, most of which are classed as asteroids and comets. Asteroids are small, mostly rocky bodies made from debris left over after the formation of the inner solar system. They originated in the region between the orbits of Mars and Jupiter (where Jupiter's gravity disrupted the formation of larger bodies) and this is where the main asteroid belt is still found today (see page 161). However, interactions between asteroids have since scattered many into orbits that bring them into the inner solar system or send them out among the giant planets. The largest asteroid, Ceres, is also defined as a dwarf planet.

Comets, meanwhile, are small icy bodies, formed in the region where the giant planets orbit today but subsequently exiled to the distant edges of the solar system. From here, they occasionally fall back towards the Sun, where they can heat up and release gas that forms spectacular tails.

Asteroid 21 Lutetia

Centaurs and ice dwarfs

Orbiting between and beyond the outer planets, centaurs and ice dwarfs are ice-rich bodies that are related to comets. Ice dwarfs are found in a region called the Kuiper Belt, stretching from the orbit of Neptune to perhaps 100 AU from the Sun. They vary in size from fairly substantial worlds like Pluto and Eris to smaller comet-like bodies, but are thought to have stayed in roughly this region of the solar system for most of their history. The doughnut-shaped 'classical' Kuiper Belt is surrounded by a more diffuse 'Scattered Disc' whose members have more elongated and wildly tilted orbits. As the name suggests, these objects are thought to have been ejected or scattered into their present orbits by close encounters with Neptune or other giant planets. Centaurs, meanwhile, orbit between the giant planets. They probably originate in the Kuiper Belt, and spend only a relatively short time in this region of the solar system before the gravity of the planets disrupts their orbits further, perhaps kicking them out to the scattered disc, or sending them into more elliptical orbits where they become fully fledged comets.

The ice dwarf Chariklo is the first Kuiper Belt Object known to have its own ring system.

The birth of solar systems

Much of what we know about the origin of our solar system is based on observations of others that are still in the process of formation. Solar systems begin life in collapsing clouds of star-forming gas and dust called 'nebulae'. The process of collapse may be triggered by tides from passing stars, by the shockwave of a nearby supernova (exploding star), or by the stellar winds blown out by other newborn stars. Eventually, random knots within the nebula become dense enough to exert their own gravitational pull, drawing in more and more gas and dust and separating into dark clouds called 'Bok globules'.

As matter falls towards the centre of a globule, it starts to spin faster (thanks to conservation of angular momentum – the same principle that causes a pirouetting skater to spin faster when they pull their arms in) and flatten out in a broad disc. The centre of the disc becomes a dense and hot protostar, while the surrounding disc contains material that will go on to form its planetary system.

The Eagle Nebula (Messier 16) is a site of present-day star formation.

Origins of our solar system

Astronomers base their ideas about our own solar system's origins on studies of the present-day Sun, models of stellar evolution and analysis of material found in meteorites that have changed little since their formation. Hence, we can be reasonably certain that the Sun itself became a fully fledged star about 4.6 billion years ago. Analysis of the elements found within meteorites even suggests that the nebula out of which the Sun formed was enriched with material from recent supernovae – our solar system may owe its existence to the shockwave from the death of another star.

The oldest meteorites found so far, dating to 4.57 billion years ago, show that solid material was condensing around the Sun at this time. Competing theories differ over just how long the planets took to come together, with estimates ranging from just a few million to around 100 million years. However, studies of rocks from the Moon and Mars suggest these worlds had formed by 4.50 billion years ago at the latest.

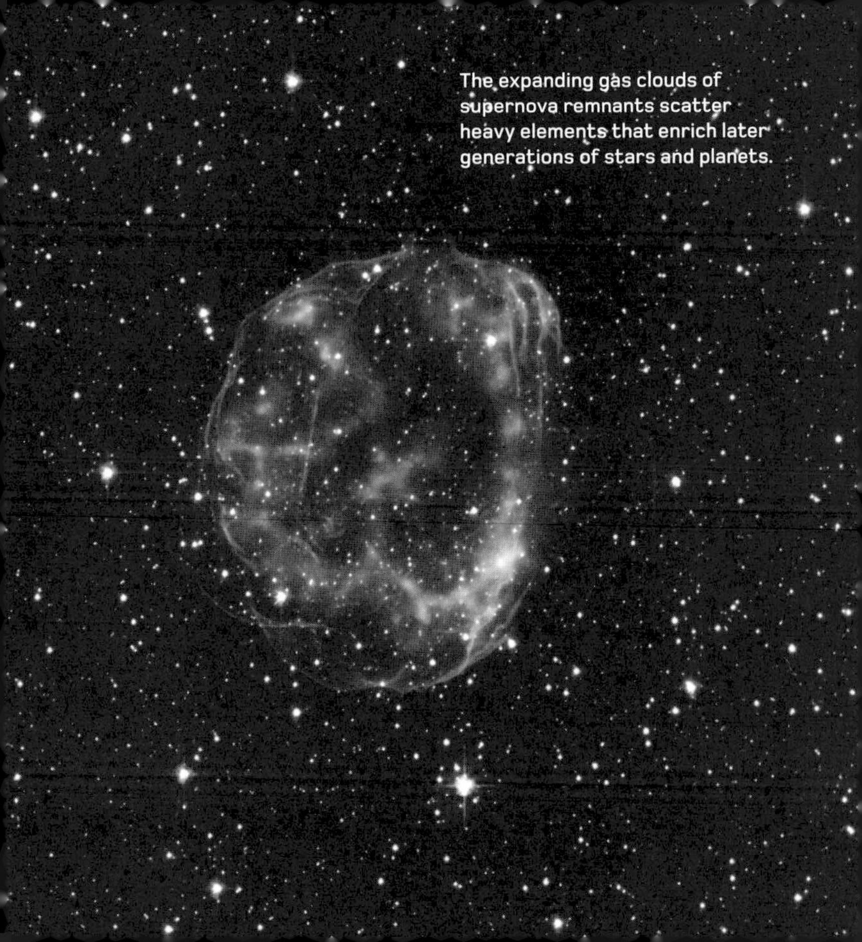

The expanding gas clouds of supernova remnants scatter heavy elements that enrich later generations of stars and planets.

Rocky planet formation

Modern ideas about planet formation are broadly based on a 'solar nebula disc model' proposed by Soviet astronomer Viktor Safronov in the 1960s. According to this model, the original 'protoplanetary' disc was a mix of gas, ice (chemicals with low melting points), and heat-resistant dust. Heat from the young Sun caused ice to evaporate, while fierce solar winds drove both gas and vapour out of the inner solar system, leaving only dust behind.

Over the next few million years, dust particles collided at random and stuck together in a process called collisional accretion. The traditional view is that some dust clumps eventually grew large enough to exert their own gravity, pulling in material from their surroundings and snowballing in size to become Moon-scale bodies called planetesimals. These eventually collided, melting and coalescing to become the planets we know today. A recent theory known as 'pebble accretion', however, argues that the planets did not grow in this piecemeal way – instead, they formed abruptly as huge drifts of orbiting dust became gravitationally unstable and underwent sudden collapse.

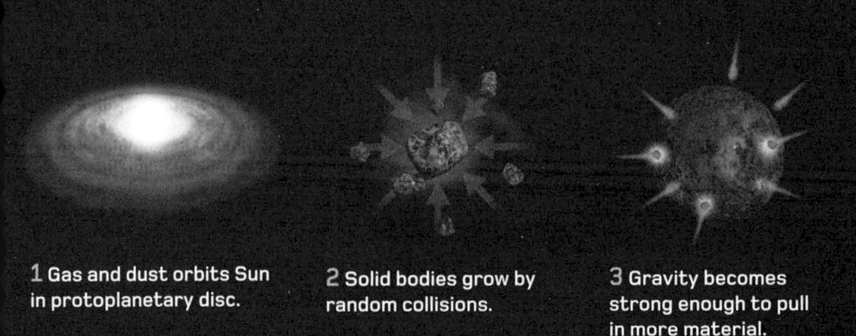

1 Gas and dust orbits Sun in protoplanetary disc.

2 Solid bodies grow by random collisions.

3 Gravity becomes strong enough to pull in more material.

4 Larger planetesimals collide to form planets.

5 Planetary surfaces are bombarded by asteroids and comets.

Formation of giant planets

Beyond the region of the present-day asteroid belt, the heat of the newborn Sun and the pressure of its solar wind were weaker. This allowed large amounts of ice to survive without melting, and for lightweight gases to linger in a vast, doughnut-shaped ring around the Sun. These materials vastly outweighed the relatively small amounts of dust in the outer solar system and, as a result, gas and ice became the raw materials of the outer planets and their moons.

One possible model for their formation mirrors that suggested for their inner neighbours. Dust and ice clumped together by chance collision, eventually forming cores with sufficient gravity to pull in a huge envelope of ice and gas. A problem with this theory lies in growing the cores quickly enough to pull gas from their surroundings before it dissipates into interstellar space. An alternative theory requires the planets to be created much more rapidly, perhaps as a result of collapsing eddies in a mostly uniform protoplanetary nebula, and to develop their structure later.

Two models of giant planet formation

Gas dominates outer part of protoplanetary nebula.

Solid cores form in the same way as rocky planets, then pull in gas from surroundings.

Large clumps of gas separate out into protoplanets which then collapse under their own gravity.

The Late Heavy Bombardment

Some time after the main era of planet formation had come to an end, worlds across the inner solar system seem to have undergone a violent bombardment from space. The oldest surfaces on planets and moons are saturated with impact craters of all sizes, caused by asteroids and comets that rained down at a rate seen neither before nor since. Younger landscapes, wiped clean by geological activity such as volcanoes, have since endured crater formation at a greatly reduced rate.

Geological dating evidence from Moon rocks suggests the so-called 'Late Heavy Bombardment' occurred between about 4.0 and 3.8 billion years ago, since this is when the vast majority of 'impact melt' rocks seem to have formed. Popular models of solar system evolution suggest that the bombardment happened as a result of disturbances among the giant planets of the outer solar system. Some scientists are sceptical about whether the bombardment 'peaked' in the way most believe, however, or whether impact rates simply tailed off steadily from the birth of the solar system.

Solar system evolution

Until recently, astronomers believed that our solar system had changed little since the end of the Late Heavy Bombardment 3.9 billion years ago. The orderly, near-circular orbits of the major planets suggested they had followed these tracks without alteration for most of their history. Since the mid-1990s, however, the discovery of solar systems around other stars and huge advances in computer modelling have painted a different picture of our system's surprisingly dynamic history. Many of these alien solar systems show very different arrangements of planets and wildly elliptical orbits, and while they are inevitably 'snapshots' of a single moment in their development, together they undermine the belief that orderly orbits are the only way for a solar system to survive in the long term. The idea that our own system has gone through traumatic changes in its past also helps to answer some otherwise puzzling features of individual planets, such as the arrangement of the gas giants, the location of the Oort Cloud and the tilted axis of Uranus.

Exchanges of orbital energy with swarms of comets may have caused the planets to shift their locations in the early solar system.

Evolving orbits

Traditional models of orbits (see page 12) treat them as an interaction between just two bodies, but the reality is far more complex – every massive body has its own gravity that effects everything else. So while a two-body model may describe an orbit at one instant, this doesn't mean the orbit will remain the same when extrapolated forwards or backwards in time.

One example of such complexities, seen across the solar system on many different scales, is 'orbital resonance'. Objects in different orbits travel at different speeds, so the distance between them (and their gravitational pull on each other) can vary hugely. Usually, this means that their influence on each other can be ignored. However, if objects happen to have 'resonant' orbits (with periods that are simple fractions or multiples of each other), they will return to the same alignments quite frequently. In this situation, the effects of even tiny gravitational tugs is magnified – over time a resonant object can pull another's orbit out of shape, or even disrupt it completely.

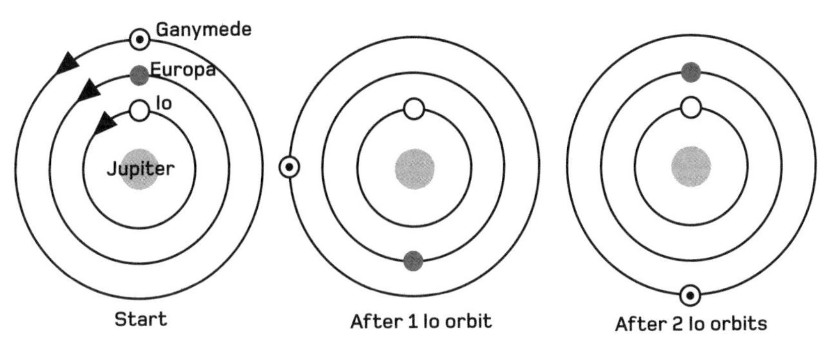

Three of Jupiter's large 'Galilean' moons display a pattern of orbital resonance, with Io orbiting the planet twice as fast as Europa and four times faster than Ganymede. Io and Europa's orbits were both driven into this arrangement long ago by the tidal tug-of-war between Ganymede, the largest moon in the solar system, and Jupiter itself.

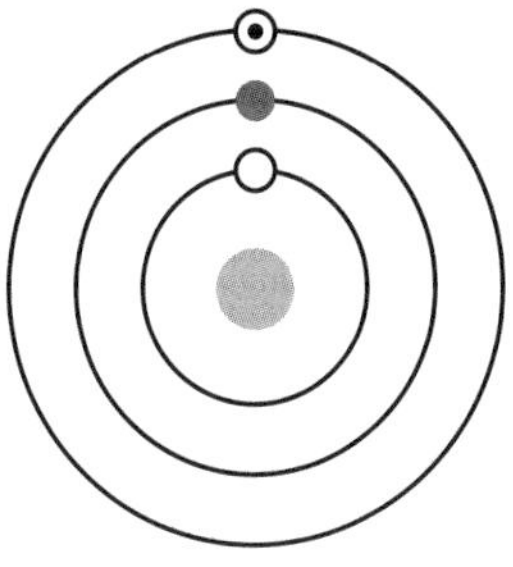

The Nice Model

One of the most influential models of orbital evolution in the solar system was developed by astronomers at France's Nice Observatory in the early 2000s. The Nice Model suggests that shortly after their formation, the four giant planets were tightly packed together, with near-circular orbits inside the current orbit of Uranus, and that Neptune, now the outermost planet, started life closer to the Sun than Uranus. Beyond the major planets lay a 'proto-Kuiper Belt' of small, icy objects. Computer simulations suggest this arrangement would have been stable for about 500 million years, before close encounters between Uranus and Neptune disrupted their orbits. At first, they moved closer to Jupiter and Saturn, but soon these giants' more powerful gravity flung the slightly smaller worlds out towards their present orbits. As they ploughed through the proto-Kuiper Belt, smaller worlds were either thrown further out (see page 366) or pushed inwards to bombard the worlds of the inner solar system. The model may even explain Uranus's sharply tilted axis (see page 291).

1 Giant planets orbit close together, surrounded by a dense 'proto-Kuiper belt'.

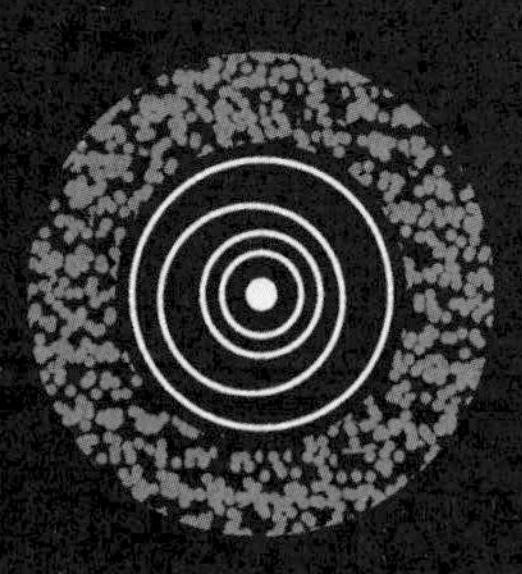

2 Jupiter and Saturn's orbits become resonant, pushing Neptune and Uranus outwards.

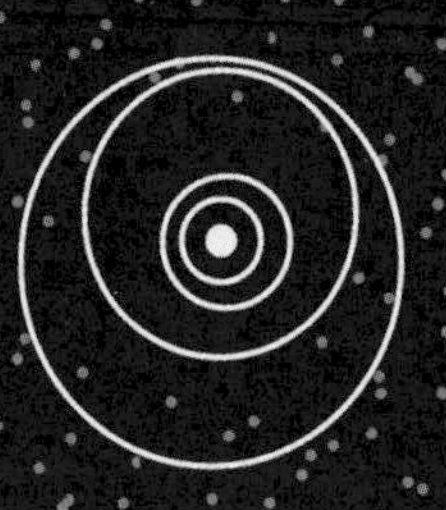

3 Ice giants plough into the proto-Kuiper belt, swapping positions as they do so.

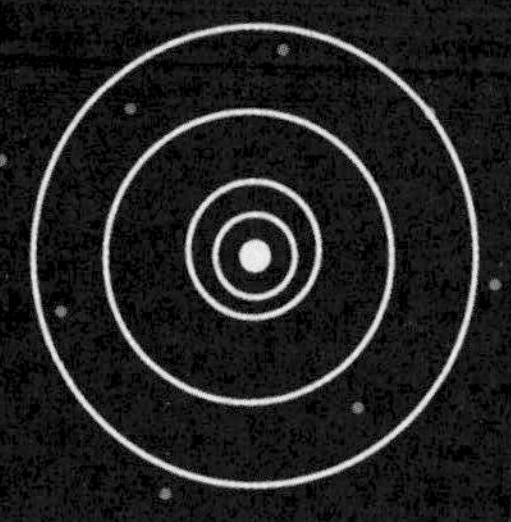

4 Planets reach their present configuration, with surviving KBOs widely scattered.

Lost planets?

Some theories of solar system evolution do more than just reshuffle the orbits of the existing planets. They suggest the existence of other planet-sized bodies that have since been lost – either crashed into the Sun, exiled to the edges of the solar system or ejected completely into interstellar space. For example, one explanation for the Late Heavy Bombardment proposes the existence of a fifth rocky planet that originally formed between Mars and the proto-asteroid belt. This world's orbit would have become unstable after a few hundred million years, sending it ploughing through the asteroids and disturbing their orbits before the planet itself plunged into the Sun.

Meanwhile, a variation on the Nice Model (see page 42) argues that the presence of a lost giant planet (an ice giant similar to Uranus and Neptune) would ultimately have made it more likely for the other worlds to settle in their current orbits. This fifth planet might also have triggered the Late Heavy Bombardment before being expelled from the solar system.

This artist's impression shows a hypothetical Neptune-like planet at the edge of our solar system.

Life beyond Earth?

A century ago, many astronomers were willing to entertain the idea of life existing on other worlds in our solar system. Discoveries over the following decades, however (culminating in the first space-probe flybys of the 1960s), revealed that our neighbouring planets Venus and Mars were far more hostile to life than had previously been suspected. In recent decades, however, the prospect of alien life on our cosmic doorstep has returned. Potentially hospitable environments have been found in surprising locations (most notably on icy moons, such as Europa and Enceladus – see pages 216 and 256), and Mars may well have reservoirs of liquid water, essential to the development of life as we know it, just beneath its surface (see page 153). Meanwhile, our understanding of life's ability to thrive in hostile conditions has been transformed by the discovery of 'extremophile' organisms on our own planet. Some astronomers have even speculated that primitive life could have started out elsewhere in the solar system and was subsequently carried to Earth by comets (the so-called 'Panspermia' theory).

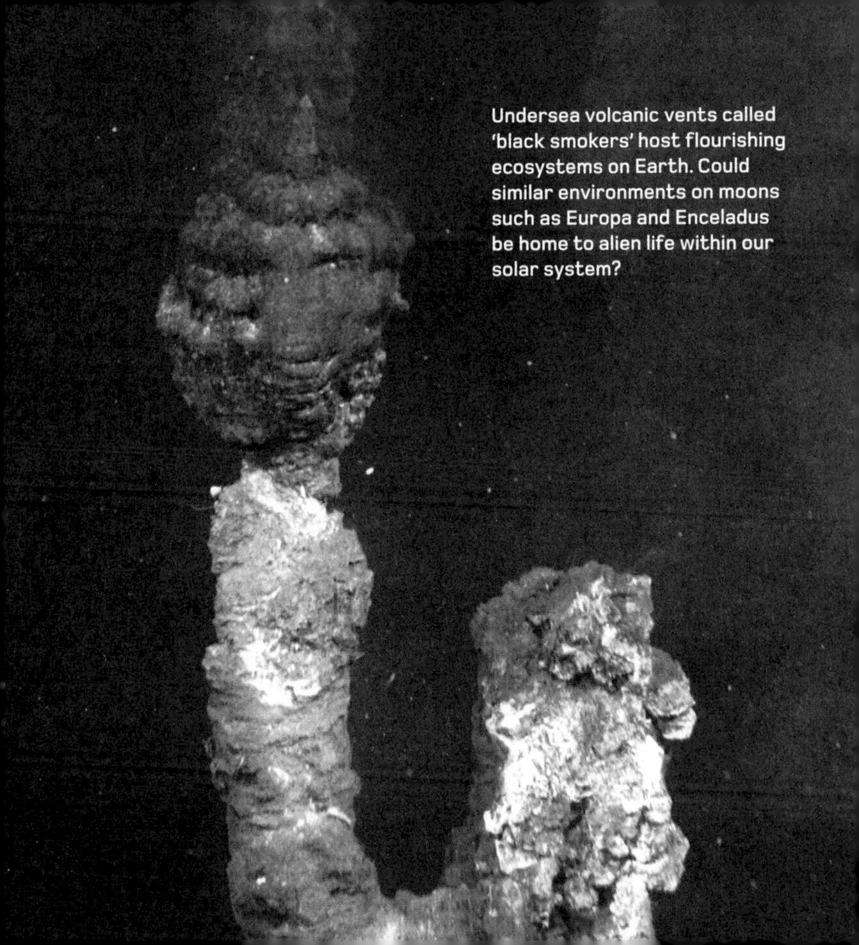

Undersea volcanic vents called 'black smokers' host flourishing ecosystems on Earth. Could similar environments on moons such as Europa and Enceladus be home to alien life within our solar system?

The future of the solar system

Mathematical limitations mean it is impossible to make predictions about the orbits of individual planets more than 50 million years beyond the present, so models of the future solar system tend to focus on the fate of the star at its centre. The Sun is currently about halfway through its 'main sequence lifetime' of about 10 billion years (during which it generates fuel by the nuclear fusion of hydrogen in its core). At the end of this stage in its life, with its principal fuel source exhausted, structural changes will see the Sun brighten considerably (perhaps by a factor of a thousand or more) and swell hugely in size, becoming a "red giant" star whose outer layers will extend beyond the orbit of Venus and perhaps engulf Earth itself.

Although we have 5 billion years until its core hydrogen runs out, the Sun will grow brighter long before that, rendering Earth uninhabitable in about a billion years. On a similar timescale, reduction in the Sun's mass due to solar wind (see page 64) is likely to see all the planets spiral slowly outwards in their orbits.

Planetary nebulae are short-lived but spectacular 'smoke rings' created when a dying Sunlike star throws off its outer layers.

Our local star

By any conceivable measure, the Sun dominates everything else in the solar system. With a diameter of 1.4 million km (865,000 miles), it is large enough to contain all objects that orbit it almost 600 times over. It also contains 99.8 per cent of the solar system's total mass.

Like all stars, the Sun is essentially a huge ball of gas, predominantly hydrogen with small amounts of helium and a few other trace elements. High temperatures strip atoms of their outer electron particles, leaving them with exposed atomic nuclei that carry electrical charges. As huge masses of gas rise and fall inside the Sun, vast electromagnetic fields are generated that affect the Sun's outward appearance (see pages 60–63) and make themselves felt across the solar system. Streams of charged particles blown out of the atmosphere form a supersonic 'solar wind' whose influence defines a vast region of space called the heliosphere, often considered to be the limit of the solar system (see page 372).

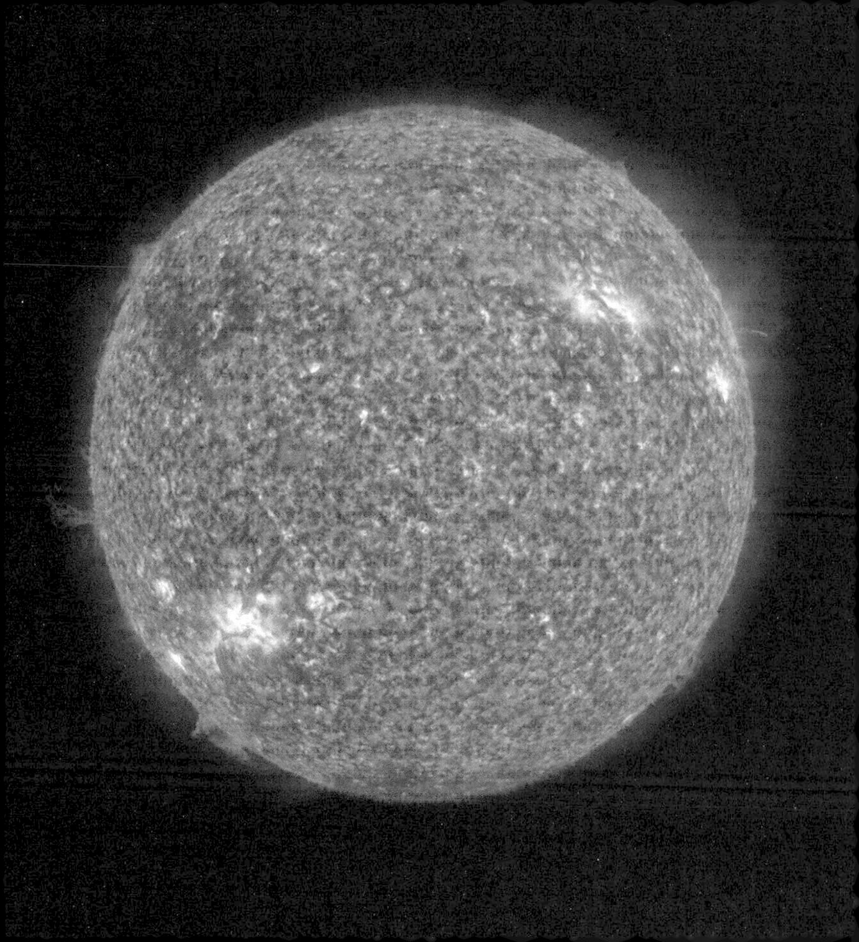

The visible Sun

From a distance, the Sun's visible surface looks like a blazing, sharply defined sphere, but appearances can be deceptive – in reality, the incandescent 'photosphere' that we consider to be the surface of the Sun is one layer among many. The photosphere appears solid because it is the region in which the Sun's gases finally become tenuous enough, and temperatures low enough (at around 5,500°C/9,900°F), for light to escape into interplanetary space. This change actually occurs across a zone about 1,000 km (600 miles) deep, so up close the photosphere would appear more like a thinning bank of fog than a solid surface. Although the Sun's surface appears uniformly bright in visible light except for occasional dark sunspots (see page 60), filtered views reveal a more complex pattern – the photosphere is covered in granular cells with bright, hot centres and cooler, darker edges. The bright cores of these planet-sized regions mark areas where hot gas from inside the Sun reaches the surface and sheds energy, while the darker edges are created where cooled gas, pushed aside, begins to sink back downwards.

Dark sunspots amid granulation patterns in the solar photosphere

Inside the Sun

The Sun's interior is broadly divided into three layers. The hot dense core is where nuclear fusion (see page 56) releases energy that gradually escapes outwards. Energy leaves the core in the form of tiny packets, or photons, of gamma rays (the most energetic form of electromagnetic radiation). It then enters the 'radiative zone', where matter is so densely packed that the rays can only travel a tiny distance before encountering a particle and bouncing in a different direction. As a result, the photons take hundreds of thousands of years to move out across this zone, slowly losing energy to their surroundings and transforming into less energetic X-rays and ultraviolet forms as they do so. At the top of the radiative zone, changes to temperature and pressure render the Sun's matter opaque. Energy can no longer be transferred as radiation, so it is absorbed, heating huge masses of gas that naturally push upwards. At the top of this 'convection zone', the Sun's gases become transparent once again, and energy can finally escape into space as visible light and other radiation.

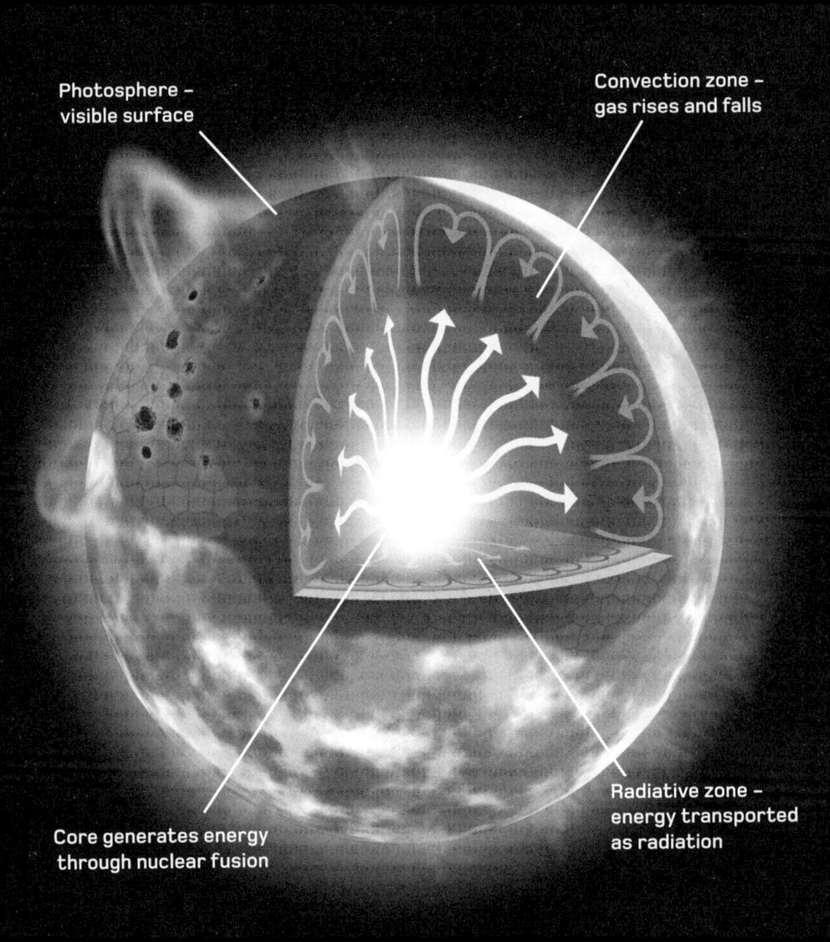
Photosphere –
visible surface
Convection zone –
gas rises and falls
Radiative zone –
energy transported
as radiation
Core generates energy
through nuclear fusion

How the Sun shines

The Sun, like all stars, is powered by nuclear fusion – a process by which the tiny central nuclei of atoms are forced together to create new, heavier nuclei. Fusion can only take place under extreme temperatures and pressures that overcome the natural tendency of nuclei to repel each other; in the Sun's case, temperatures of 15 million°C (27 million°F) and pressures 260 billion times greater than that of Earth's atmosphere. Energy released by fusion helps to heat the core and maintain these conditions of extreme heat and pressure.

Fusion in the Sun mostly involves the so-called 'proton-proton chain reaction', a process that combines protons (the simple nuclei of hydrogen, the lightest and most abundant element in the Universe) to create helium, the next lightest element. Each fusion reaction converts a tiny amount of mass directly into energy (in accordance with Einstein's equation $E=mc^2$). Owing to the Sun's huge size, however, this amounts to 4 million tonnes of material per second and an output of 3.8×10^{28} watts.

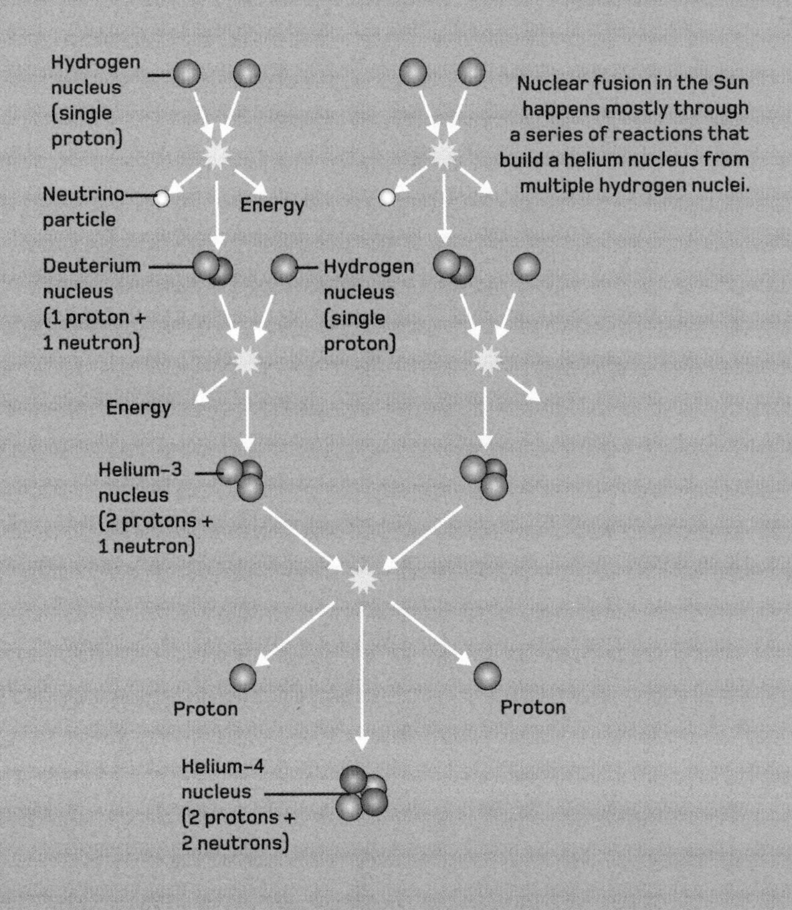

Hydrogen nucleus (single proton)
Neutrino particle
Energy
Nuclear fusion in the Sun happens mostly through a series of reactions that build a helium nucleus from multiple hydrogen nuclei.
Deuterium nucleus (1 proton + 1 neutron)
Hydrogen nucleus (single proton)
Energy
Helium-3 nucleus (2 protons + 1 neutron)
Proton
Proton
Helium-4 nucleus (2 protons + 2 neutrons)

The Sun's atmosphere

Above the Sun's brilliant photosphere lie several other layers that are sometimes collectively called the solar atmosphere. The innermost of these layers, known as the chromosphere, is around 5,000 km (3,100 miles) deep. It takes its name from its reddish colour (visible only during eclipses), and is the location of bright, hot plumes of gas known as prominences, and long, flame-like spicules that extend into the layers above. Temperatures in the chromosphere are significantly hotter than in the underlying photosphere, rising to around 35,000 °C (63,000°F) at the top. The source of energy that powers this heating is still poorly understood, but is probably associated with the Sun's magnetic field. Above the chromosphere lies a thin transition region just a few hundred kilometres deep, where changes to sparse helium gas atoms cause them to absorb radiation. As a result, temperatures in this region soar to 1 million °C (1.8 million °F) or more. This heating effect continues into the Sun's outer atmosphere or corona – a vast extended halo of thin gas that is shaped by the Sun's magnetism and, ultimately, blends into the solar wind (see page 64).

During a total solar eclipse, the Sun's tenuous outer atmosphere can be seen with the naked eye.

Solar activity

While some of the Sun's surface features, such as granulation and spicules, are permanent, others come and go. Most obvious of these are sunspots – dark patches in the photosphere that may be many thousands of kilometres across, and can persist for days or weeks. The spots appear dark because the material in them is cooler than their surroundings (though still at temperatures around 3,000°C or 5,400°F). They form in pairs, where loops of the Sun's magnetic field push out through the photosphere and create areas of lower-density gas. These magnetic loops are also linked to other forms of solar activity. Gas flowing along them forms glowing loops called 'prominences', which can be seen above the Sun's edge during solar eclipses. Furthermore, when the magnetic loops become overextended, they can 'short circuit', reconnecting closer to the surface and releasing a short-lived but huge burst of energy called a solar flare. The flare heats the surrounding solar atmosphere to millions of degrees and ejects energetic particles into space at speeds of up to 1,000 km (620 miles) per second.

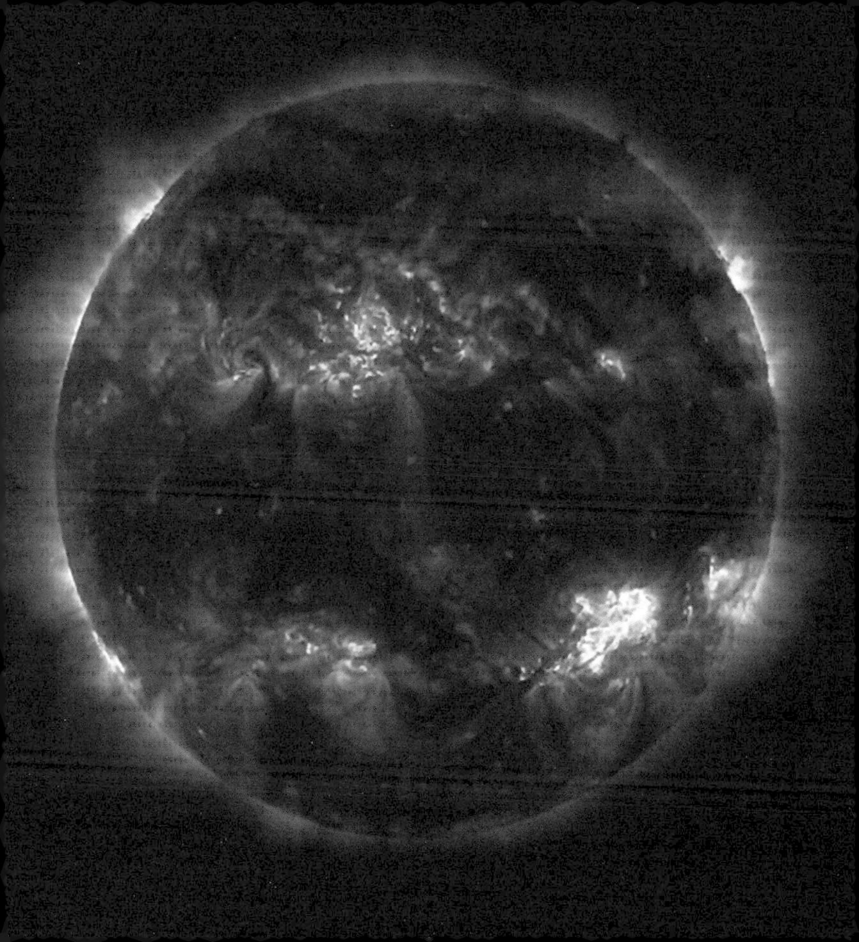

The solar cycle

Although the Sun's overall energy output is fairly steady, it is not immune to short-term changes. Sunspot numbers and solar flares vary in number and intensity according to a solar cycle that repeats roughly every 11 years. The cycle begins with a relatively placid Sun and a few sunspots at high latitudes, then intensifies as sunspots increase and move towards the equator. Finally, the cycle subsides as sunspots nearing the equator disappear.

The solar cycle is driven by changes in the Sun's magnetic field, which is generated not in the core, but by swirling gas in the convection zone. Because the Sun's interior rotates more slowly at higher latitudes, the magnetic field becomes tangled over time, with loops erupting through the surface to create active regions of sunspots and flares, and eventually cancelling out as they start to make connections across the equator. The entire magnetic field regenerates with each solar cycle (with the magnetic poles 'flipping' each time), but the causes of deeper, decades-long variations in levels of activity are still poorly understood.

1 Cycle begins with magnetic field aligned from pole to pole.

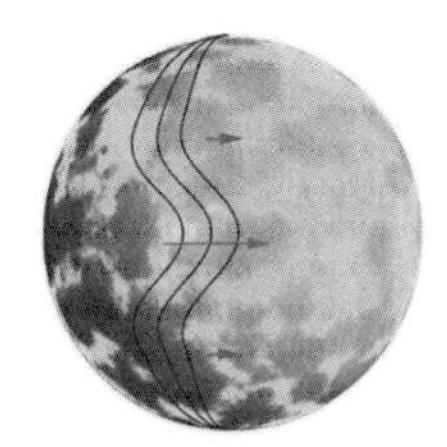

2 Over time, faster rotation of the Sun's equator distorts the field.

3 Magnetic loops push out through photosphere, creating sunspots and prominences.

4 Loops emerge closer to equator as field is twisted further.

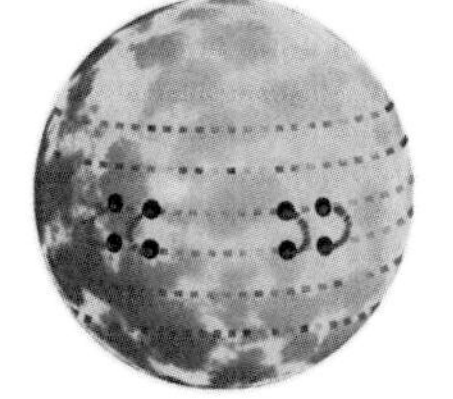

5 As loops approach the equator they begin to cancel out, weakening the magnetic field.

6 At the end of an 11-year cycle, the field regenerates with the opposite orientation.

Solar wind

High temperatures and strong magnetic fields in the Sun's corona boost the energy of its electrically charged particles. This allows them to overcome the Sun's strong gravitational field and escape in a wind that is felt across the solar system. Solar wind particles are mostly fragments of the lightweight gaseous elements that dominate the Sun – atomic nuclei of hydrogen and helium, and electrons (the tiny particles that usually orbit the nucleus to complete an atom, but that are stripped away by intense coronal temperatures). Close to the Sun, the particles travel in a smooth flow at supersonic speeds, with faster wind typically emerging from the poles than the magnetically tangled equator. Solar flares and even larger events called coronal mass ejections (CMEs) – caused when loops of the Sun's magnetic field 'reconnect' at lower levels and liberate huge amounts of energy – can eject vast clouds of material into the solar wind. As these clouds sweep past the planets in the following hours or days, they can distort planetary magnetic fields and create brilliant light shows known as auroral storms (see page 110).

The heliosphere

The flow of the solar wind defines a region where the Sun's influence is dominant, known as the heliosphere. Particles escape the Sun at supersonic speeds of hundreds of kilometres per second, blowing out a 'bubble' in the surrounding interstellar medium (the sparse but nevertheless measurable scattering of potentially star-forming gas and dust in the plane of our galaxy). However, the medium in its turn exerts a pressure that pushes back at the solar wind, especially in the direction of the solar system's motion.

Between 75 and 90 AU from the Sun, at a boundary called the termination shock, the solar wind's speed falls to less than the speed of sound as it passes through (about 100 km/60 miles per second). Beyond this, in a region called the heliosheath, the wind's hitherto smooth flow becomes turbulent. Finally, at the heliopause (crossed by NASA's Voyager 1 spaceprobe in 2012 at about 121 AU from the Sun), the wind's outward drift comes to a halt in the face of pressure from the interstellar medium.

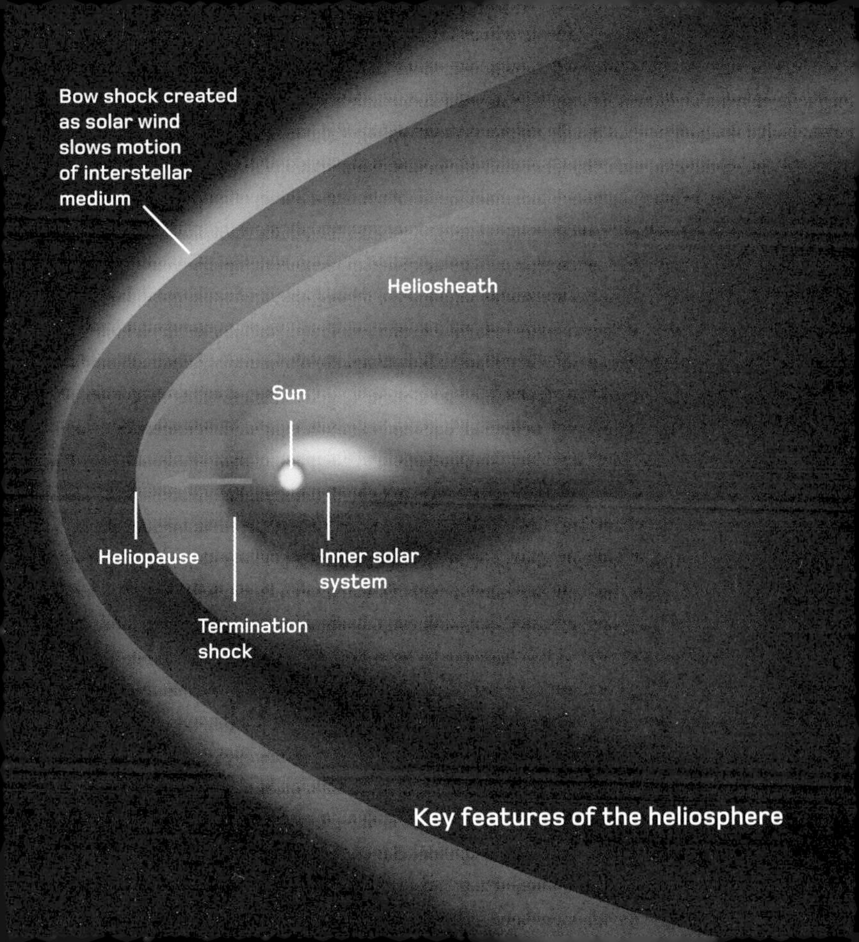
Bow shock created as solar wind slows motion of interstellar medium
Heliosheath
Sun
Heliopause
Inner solar system
Termination shock
Key features of the heliosphere

Mercury

The smallest major planet and the closest to the Sun, Mercury hurtles around its markedly elliptical orbit in just 88 days. Around 40 per cent bigger than Earth's Moon, Mercury is a similarly barren grey world, pockmarked with impact craters from the early days of the solar system, but also showing traces of ancient volcanic activity and an unusual internal structure. Atoms blasted out of surface rocks by fierce solar radiation form a sparse atmosphere, barely worthy of the name, that has to be constantly replenished as its fast-moving particles blow away into interplanetary space.

It's predictable that Mercury has one of the hottest surfaces in the solar system, reaching temperatures of up to 425°C (800°F) at midday, but it is perhaps surprising that it's dark side can be one of the coldest, plunging to –195°C (–319°F). This is because Mercury has a unique relationship between its day and its year that gives it some of the longest nights in the solar system (see page 70).

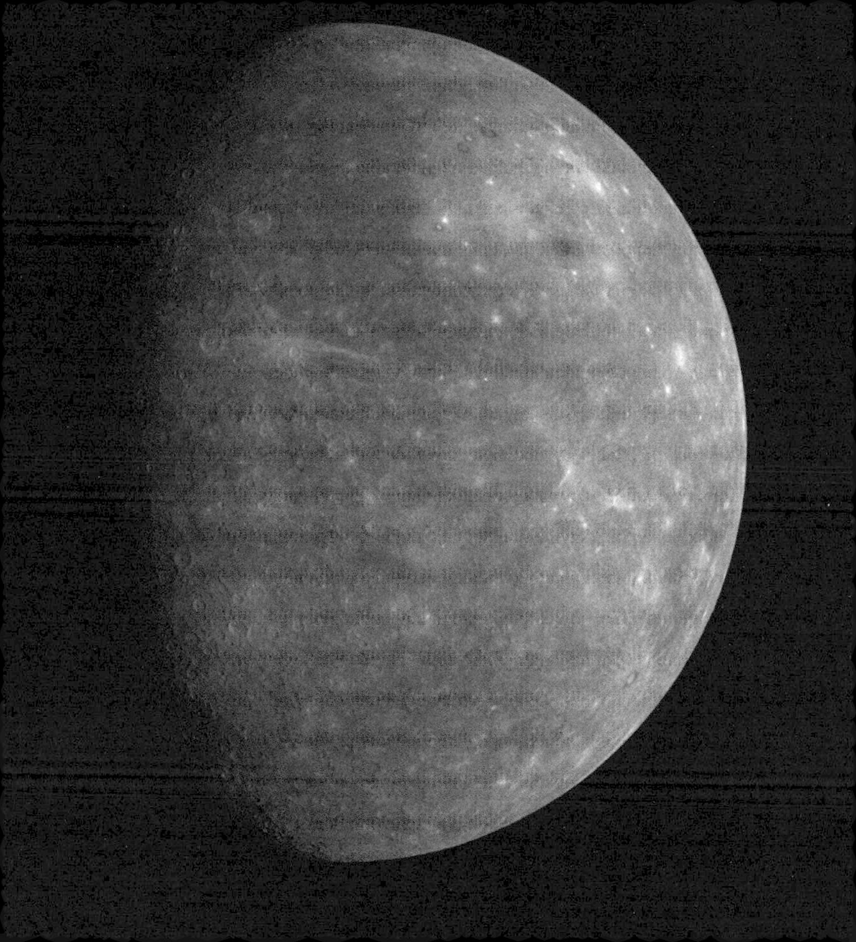

Orbit of Mercury

Mercury's orbit is the most elliptical of any major planet; its distance from the Sun ranges from 0.31 to 0.47 AU. Tides caused by gravity pulling the planet's interior out of shape have long since forced Mercury to develop a 'resonant' rotation period that minimizes tidal forces. As a result, Mercury spins on its axis with a period of roughly 57 Earth days that is precisely two-thirds of its 88-day year. This means that the Sun moves very slowly across Mercury's skies, with an average interval of two Mercury years between successive sunrises. However, when Mercury is at its closest to the Sun and moving at its fastest speed along its orbit (about 57 km/35 miles per second), the effects of Mercury's orbit can actually outpace its daily rotation, causing the Sun to move backwards on its path across the sky and, at the right locations, even rise, set and rise again in rapid succession. Tidal forces have also eliminated any trace of axial tilt, so Mercury orbits the Sun 'bolt upright'. As a result, the Sun barely skims the horizon at its poles, allowing permanently shadowed craters to retain deep reservoirs of ice.

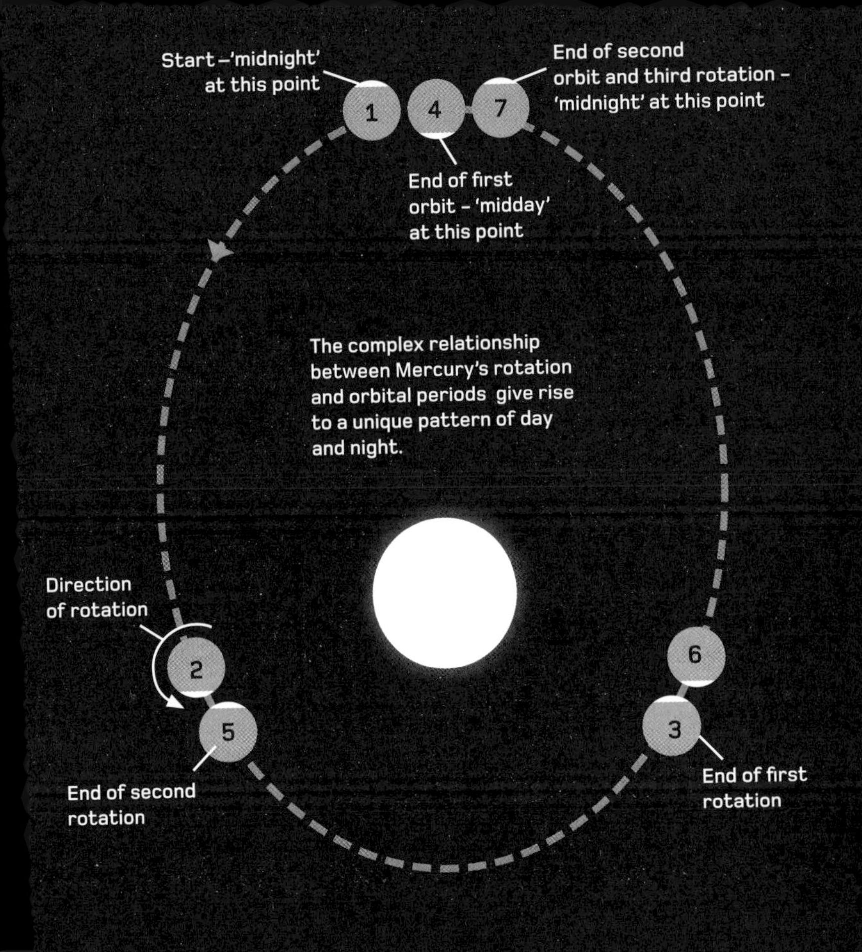
Start –'midnight' at this point
End of second orbit and third rotation – 'midnight' at this point
1
4
7
End of first orbit – 'midday' at this point
The complex relationship between Mercury's rotation and orbital periods give rise to a unique pattern of day and night.
Direction of rotation
2
5
6
3
End of second rotation
End of first rotation

Interior of Mercury

As a rule of thumb, astronomers expect smaller solar system bodies to be less dense than larger ones made of similar materials – simply because the smaller worlds have less gravity to compress their interiors. It was somewhat unexpected, then, when the Mariner 10 flybys of 1974/75 revealed that Mercury is almost as dense as Earth.

The reason for Mercury's high density is the unusually large metallic core that occupies 55 per cent of the planet's interior. The core's size is thought to be the result of a cataclysmic collision early in Mercury's history; the theory is that Mercury was once significantly larger, before much of its mantle was blasted away into space by a glancing impact. The outsized core had a significant effect on Mercury's later evolution – it has retained enough heat to stay partially molten to the present day, and swirling electric currents within its liquid layer generate a magnetic field around the planet. Changes to the core have also affected surface features (see page 76).

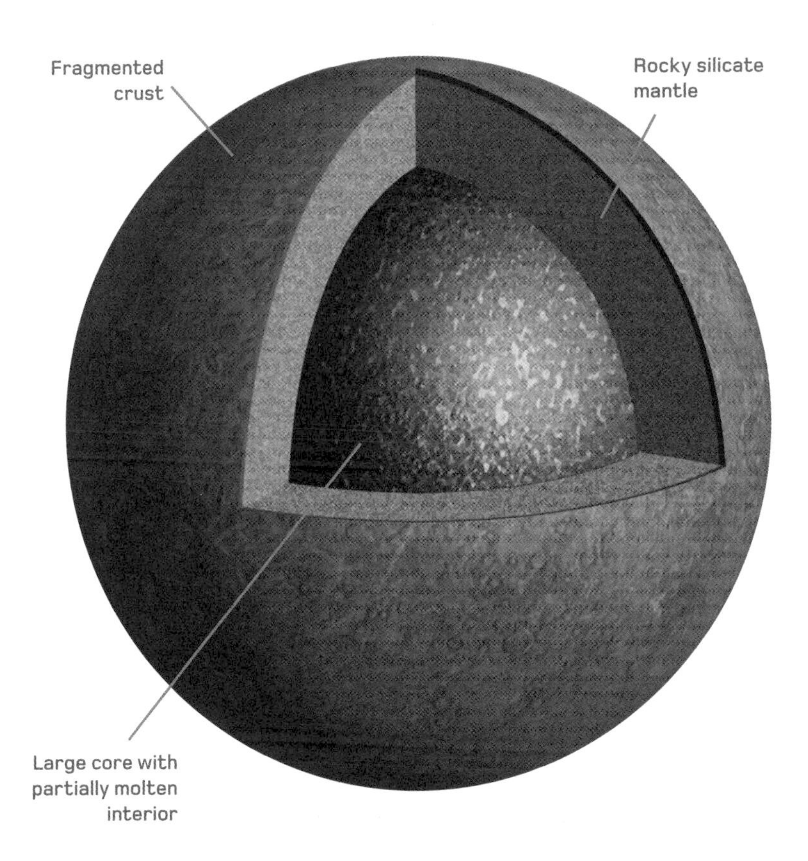
Fragmented crust
Rocky silicate mantle
Large core with partially molten interior

Caloris Basin

Mercury's proximity to the Sun and its lack of atmosphere have seen it accumulate countless craters over its 4.5 billion years of history, which the planet's limited geological activity has done little to erase. The largest craters are known as impact basins, and at 1,550 km (960 miles) across, the Caloris Basin is the largest known impact structure in the entire solar system.

Caloris formed when a large asteroid struck the planet about 3.8 billion years ago. Rimmed by a triple ring of mountains up to 2 km (1.2 miles) tall, its interior is filled with flat lava plains – the result of volcanic activity triggered by the impact that subsequently flooded the surface. A curious crater surrounded by radiating troughs (known appropriately as the Spider) may mark the origin of some of the volcanic lava. The Caloris impact was so large that it sent seismic shock waves rippling out through Mercury's crust and straight through its core. Where they rejoined on the opposite side of the planet, they created a jumbled, chaotic region of so-called 'weird terrain'.

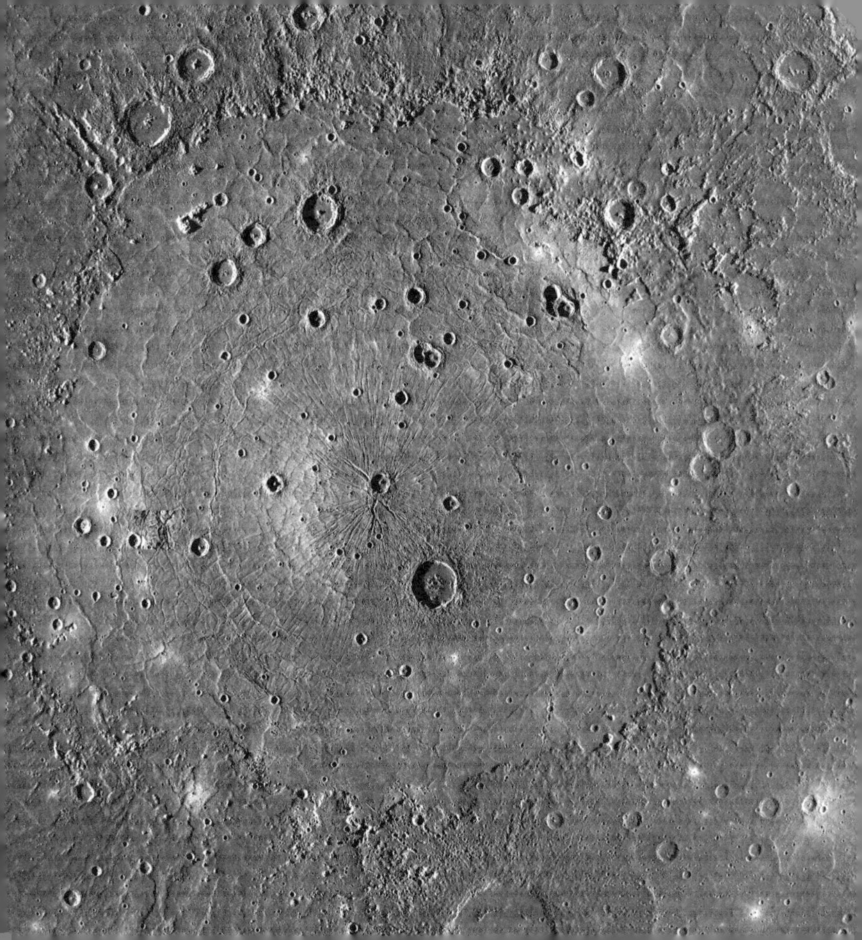

Scarps and rupes

Mercury's most unusual and unique surface features are long, clifflike escarpments and troughs known as 'rupes' that run for hundreds of kilometres across its surface, often straight through older features. For example some craters are split down the middle, with one half sitting on a raised plateau up to 1 km (0.6 miles) or so higher than its counterpart. The overall impression is that Mercury's crust has been split into a jigsaw that does not quite fit together, so some 'crustal units' are forced to sit higher than the others.

Astronomers attribute this unusual situation to Mercury's oversized core, which occupies more than half of the planet's interior. Models of the core's thermal history suggest that shortly after Mercury's formation, the core heated and swelled, causing the overlying crust to split apart. Since then, the core has cooled and shrunk in size. As the now-outsized units of crust have gradually fallen back, they have been jammed together with some forced to sit higher than others.

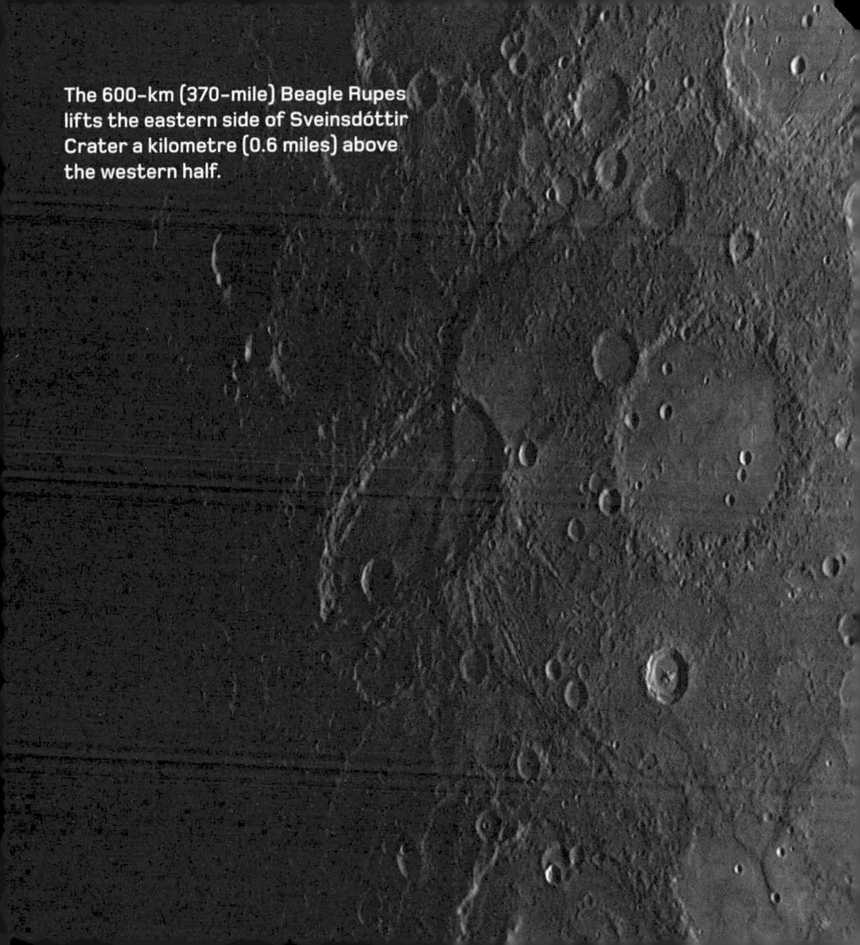

The 600-km (370-mile) Beagle Rupes lifts the eastern side of Sveinsdóttir Crater a kilometre (0.6 miles) above the western half.

Ancient volcanoes

Although Mercury's small size should have led it to cool much more rapidly than the larger rocky planets, its large core seems to have delayed that process, heating the mantle and powering geological activity for much of Mercury's history. Mercury never developed the kind of plate tectonics found on Earth, but there is plenty of evidence for widespread volcanism. Smooth plains resemble the maria, or seas, of solidified lava seen on Earth's Moon – many form a loose ring around the Caloris Basin. The fact that they overlap ejecta from that basin's formation shows they formed more recently, perhaps up to 3.5 billion years ago. Elsewhere, many impact craters have sunken pits on their floors and are covered with depositions of distinctively coloured rocks. Scientists think these are probably the result of underground magma chambers giving way beneath the crater floor. Small bright patches of material, meanwhile, resemble those released by explosive volcanism on Earth – evidence suggests these could have continued until as recently as 1 billion years ago.

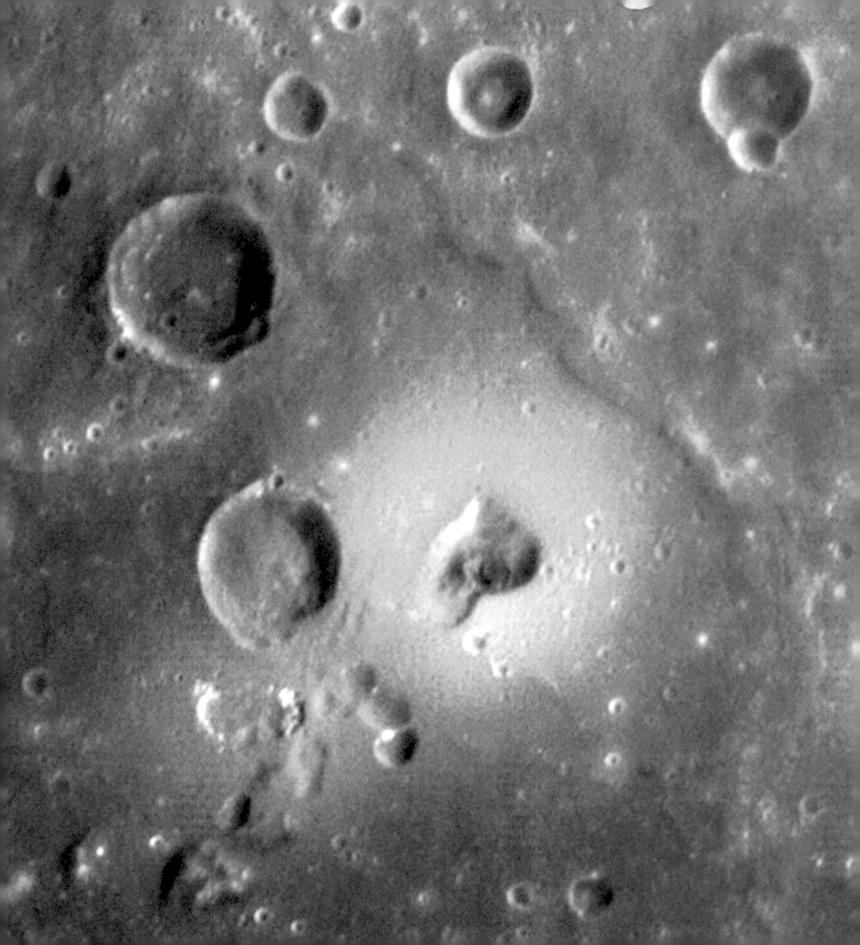

Venus

After the Sun and Moon, Venus is usually the brightest object in Earth's skies. It is thanks to its brilliance that it is named after the goddess of love and beauty in ancient Roman mythology. In terms of size, Venus is a near-twin to our own planet, just a little smaller (with a diameter of 12,104 km or 7520 miles) and orbiting a little closer to the Sun at 0.72 AU. This led some 19th- and early 20th-century astronomers to speculate that Venus might be a haven for alien life with a climate not too different from Earth's tropics.

But Venus's beautiful name is deceptive – in reality, the second planet from the Sun is a hellish furnace of a world blanketed by a choking, toxic atmosphere. Robot probes attempting to land on the surface (see page 382) fail after a few minutes at most, thanks to a combination of acid rains, searing temperatures and crushing pressures. It's only thanks to orbiting probes equipped with cloud-piercing radar that we now have an understanding of Venus as a world shaped primarily by volcanic forces.

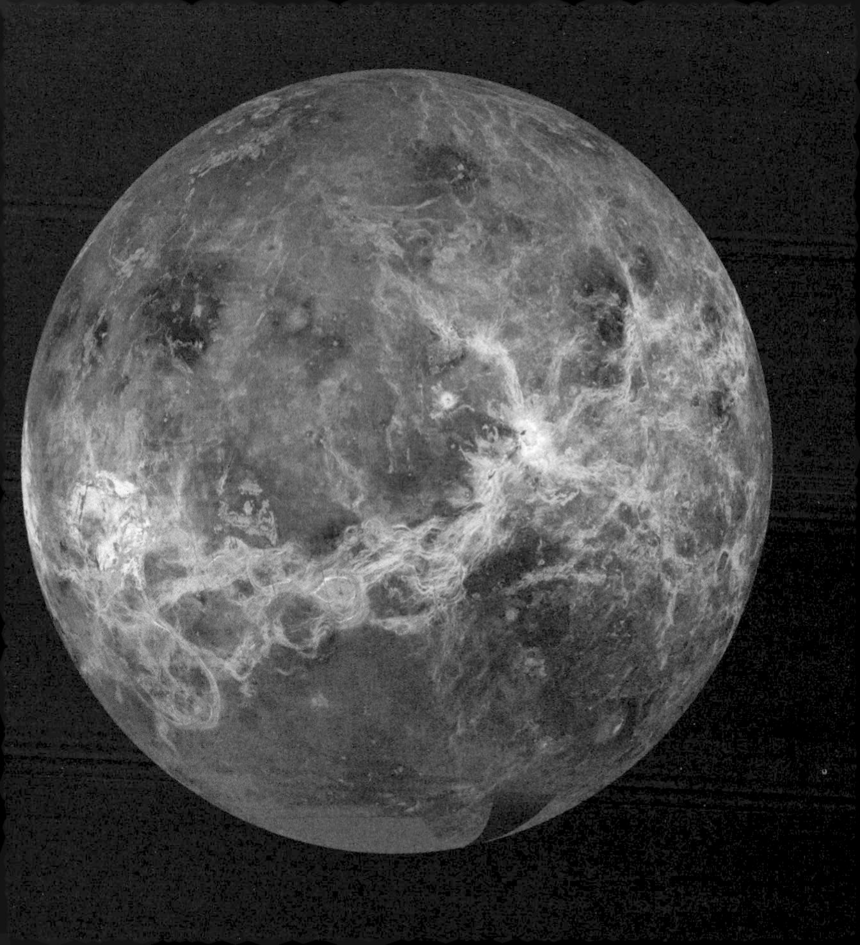

Interior of Venus

The challenge of placing instruments on the Venusian surface means that most of our information about the planet's internal structure comes from comparison with Earth rather than direct measurement. Venus's slightly smaller size and lower mass mean it is slightly less dense than Earth, but it probably has a similar elemental composition with a deep silicate crust and a nickel-iron core. The planet's smaller size means it should have cooled slightly more rapidly than Earth, but it should still have a liquid outer core. However, the lack of a magnetic field around Venus suggests something is different in the core, which seems to lack the swirling currents of molten metal which should create an electromagnetic dynamo effect. Scientists suspect the difference lies in a lack of convection (bulk movement of hot liquid metal up or down within the core) – it seems that something has suppressed the 'normal' mechanisms of heat transfer familiar from Earth. This may be linked to the lack of tectonic activity allowing heat to escape at the surface (see page 86).

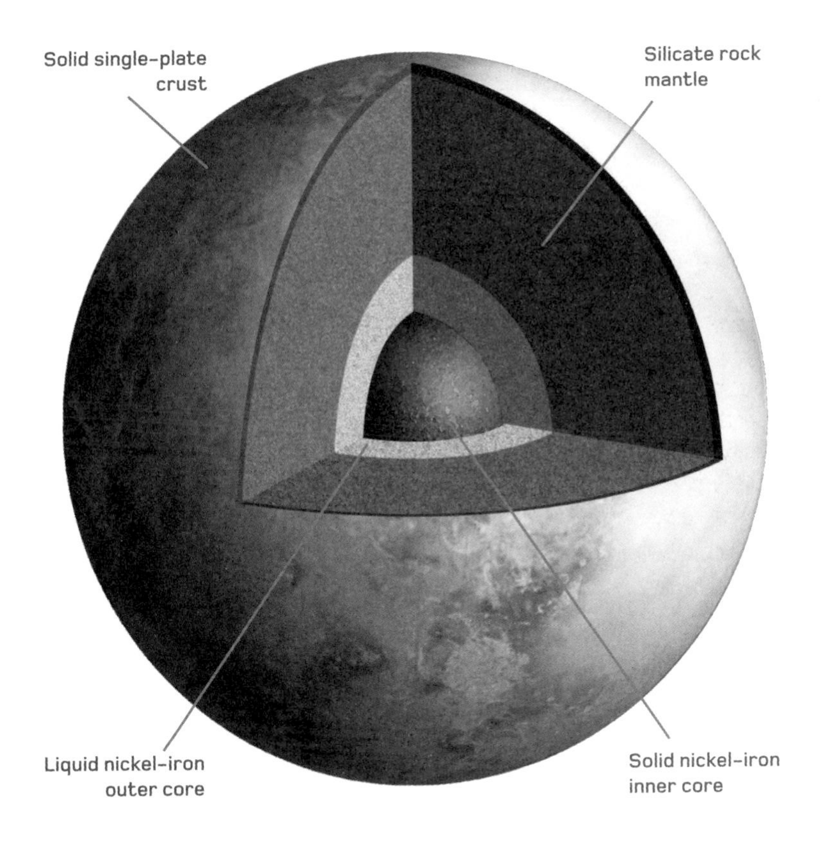
Solid single-plate crust
Silicate rock mantle
Liquid nickel-iron outer core
Solid nickel-iron inner core

The Venusian atmosphere

Venus has a uniquely hostile atmosphere, with a pressure more than 100 times higher than Earth's, temperatures of around 460°C (860°F) and sulfuric acid rain that falls from high clouds but evaporates closer to the ground. The atmosphere is dominated by carbon dioxide, generating a runaway greenhouse effect that explains the planet's soaring temperatures. Seen from space, Venusian clouds reflect more than three-quarters of the sunlight shining onto them, explaining the planet's dazzling, featureless appearance in visible light. Ultraviolet cameras reveal vast chevron-shaped cloud patterns that circle the planet in about four days.

Friction with the thick atmosphere is thought to be linked to Venus's unique rotation. Over time, the planet's spin has slowed to such an extent that it now rotates once every 243 Earth days, in the opposite direction to other planets. This means that the Venusian day is longer than its year; the Sun rises in the west and tracks slowly across the sky before setting in the east.

Venusian volcanoes

Radar maps of Venus reveal a world shaped largely by volcanic activity. Vast lava plains stretch between huge conical and shieldlike volcanoes, and some volcano types are unique in the solar system (see page 90). Estimates from counts of impact craters (see page 88) suggest that a phase of major volcanic activity came to an end 500 million years ago. Some areas may still be active – a number of volcanic peaks appear hotter than their surroundings in infrared images, and descending probes recorded lightning storms over others, similar to those seen on Earth.

Radar also shows that Venus's crust is not broken into Earth-like tectonic plates, whose boundaries act as a focus for long-term volcanic activity. Many scientists think this *lack* of tectonics is the cause of Venus's widespread volcanism. The theory is that Earth's tectonics, lubricated by ocean water, allow the slow but steady release of heat from the interior. Venus's solid crust, in contrast, acts like a pressure cooker, trapping heat for millions of years until it eventually escapes in a wave of planet-wide eruptions.

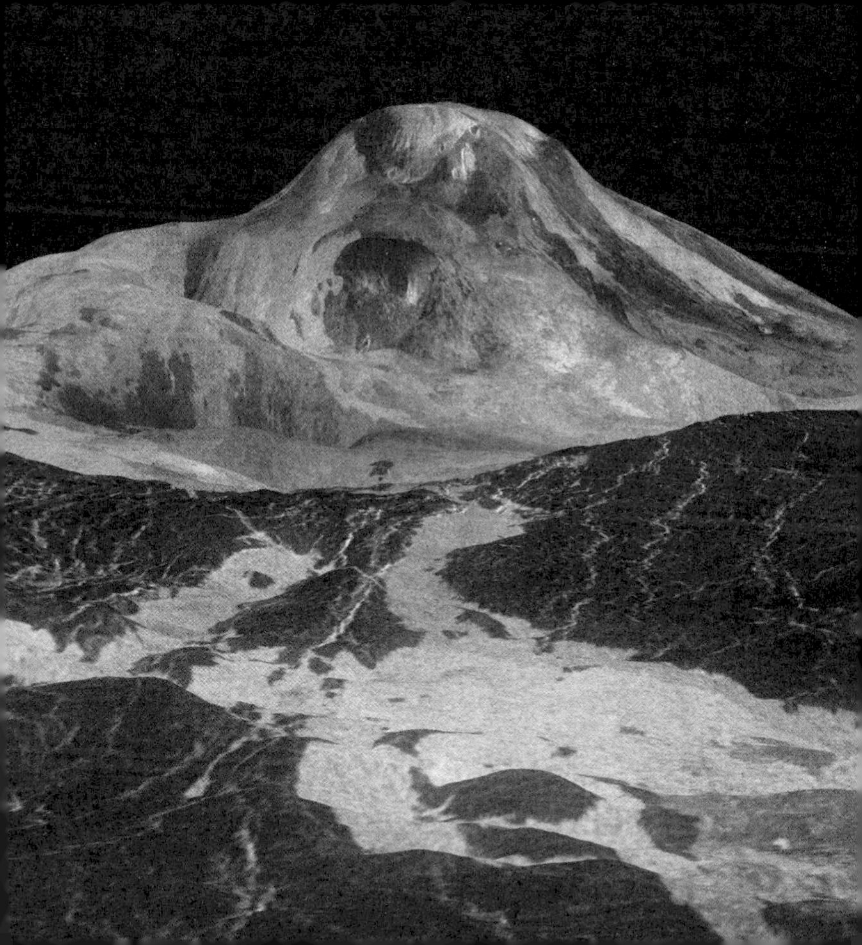

Venusian craters

Craters on Venus are rare for two reasons. First, and most obviously, the planet's dense atmosphere ensures that most incoming meteors burn up completely before they can reach the ground. Secondly, the Venusian landscape has been repeatedly resurfaced by widespread volcanism, ensuring that most traces of older impacts have been wiped away. Only a few highland areas were left unaffected by the last major resurfacing that came to an end about 500 million years ago, so craters found elsewhere must have formed since.

Because incoming meteorites must be a certain size to reach the surface, even the smallest Venusian craters are at least 30 km (19 miles) across. Another notable effect of the atmosphere is to break up large objects during descent, resulting in crater clusters such as the Howe, Danilova and Aglaonice craters shown opposite. Dense air also limits the escape of the ejecta material thrown out during impact, creating compact 'splashes' beyond the crater walls.

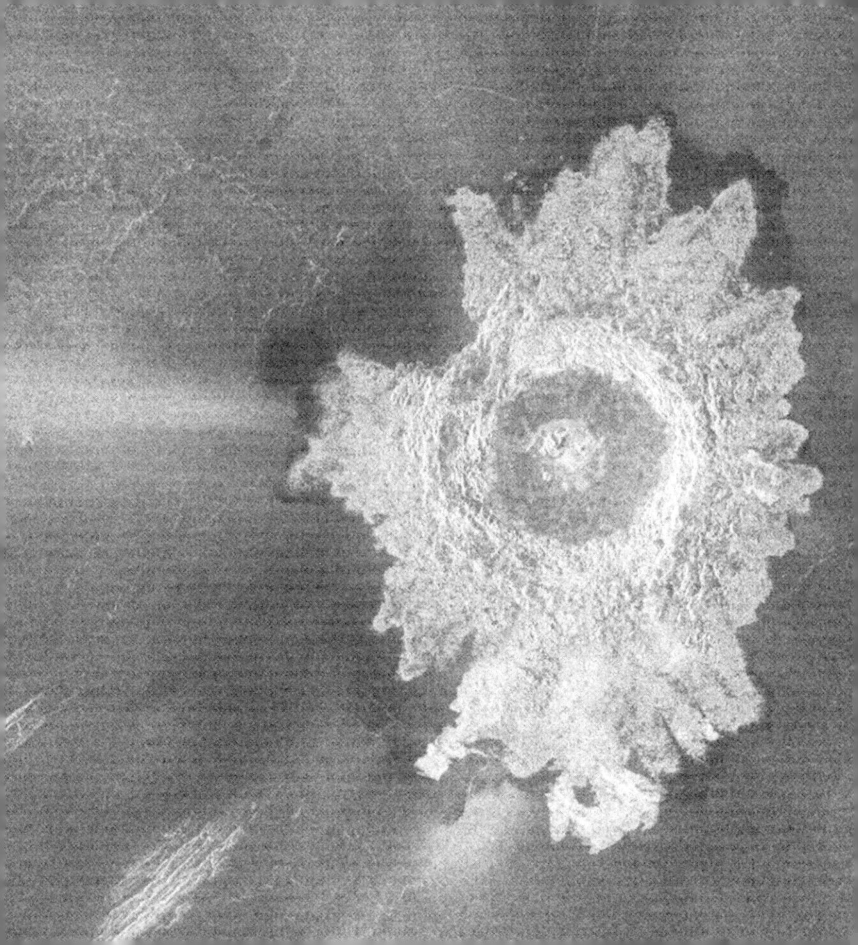

Pancake domes, coronae and arachnoid domes

Several volcanic structures are found nowhere else in the solar system. Pancake domes are flat-topped, steep-sided 'pillows' of volcanic rock tens of kilometres wide, formed as viscous lava erupted through fissures in the surface and solidified before it could spread over the surrounding terrain.

Coronae (such as Aine Corona, opposite), in contrast, are slightly depressed circular regions, typically several hundred kilometres wide, surrounded by concentric rings of cracks. They seem to mark regions where underlying molten magma once pushed the surface upwards into a dome; when the magma pressure later withdrew (perhaps drained by eruptions elsewhere), the dome subsided to form a corona. Finally, arachnoids are marked by spiderweb-like networks of cracks in the surface. Like coronae, they are believed to mark regions forced upwards by magma from below, except in this case, it's thought that the pressure grew so great that lava eventually erupted to the surface through a radial network of cracks.

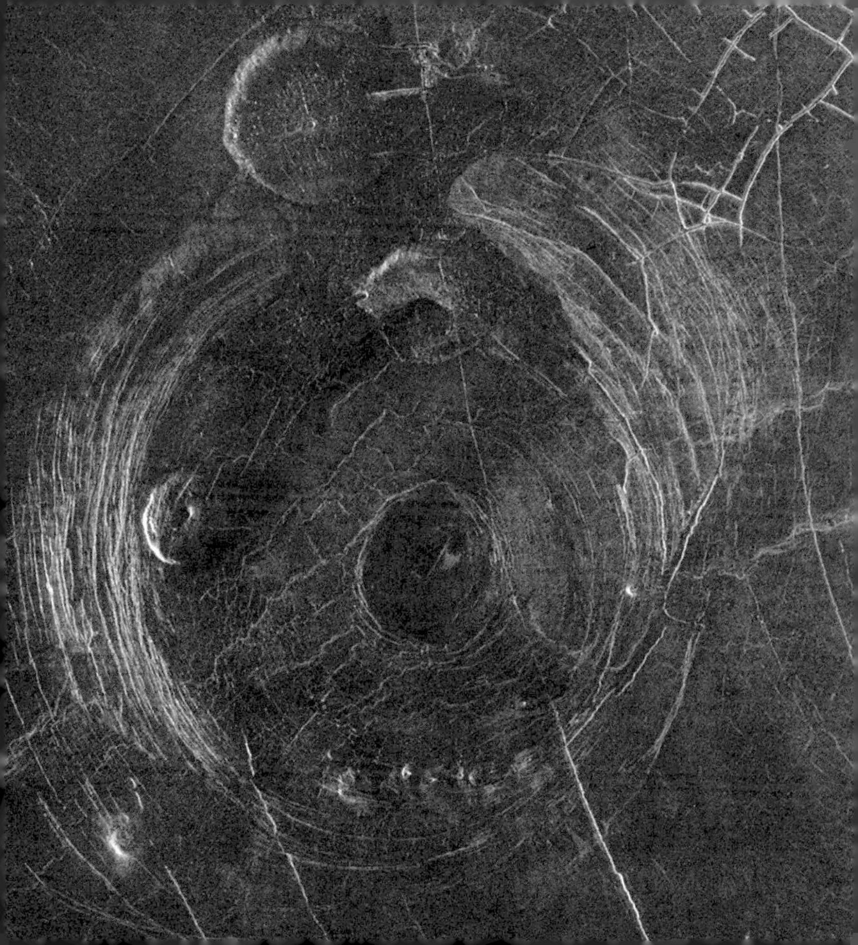

Earth

Our home planet is the third in order from the Sun. With a diameter of 12,756 km (7926 miles), it is the largest of all the solar system's rocky worlds. When compared to the other rocky planets, Earth's most distinctive features are the fact that its crust is split into slow-moving tectonic plates, the presence of abundant surface water and, of course, copious life.

All three of these phenomena are due in part to our planet's location in the solar system. Earth's near-circular orbit sits neatly within the solar system's so-called 'Goldilocks zone' – the region around the Sun where temperatures are neither so hot that surface water boils into the atmosphere and is lost to space, nor so cold that it freezes permanently. Deep oceans of water that collect in surface basins both lubricate the tectonic plates and allow for the evolution and survival of life. Interactions between tectonics, life, oceans and atmosphere give Earth the most complex environment in the solar system.

The early Earth

Our planet is thought to have formed shortly after the birth of the Sun, around 4.54 billion years ago. Early in its history, a cataclysmic collision with a Mars-sized primordial world is thought to have given birth to Earth's Moon (see page 126), and also added new material to Earth itself. As the crust began to solidify, Earth rapidly developed an early atmosphere rich in hydrogen and water vapour, likely through a mix of 'outgassing' from volcanic eruptions and bombardment with ice-laden comets and asteroids. As the Earth cooled, water began to condense into oceans.

The cores of the continents are thought to have grown fairly rapidly – early in Earth's history, the escape of heat from the interior would have created new crust and driven tectonic plates into one another at an accelerated rate. This powered head-on collisions that built up thick regions of crust, called 'cratons', over hundreds of millions of years. Traces of 'biogenic' chemicals linked to the activity of primitive life suggest that the first communities of microbes were established by at least 3.7 billion years ago.

A simulated view models the effects of a large asteroid impact on the crust of the young Earth.

Earth in its orbit

Our planet orbits the Sun at an average distance of 149.6 million km (93 million miles), taking almost exactly 365.25 days to complete one orbit. As a result, from our point of view on Earth, the Sun appears to move eastwards around the sky relative to the more distant stars, returning to exactly the same position one year later. At the same time, Earth spins on its axis in about 23 hours and 56 minutes, causing the Sun and stars to apparently move westward across the heavens. Our 24-hour day, therefore, accounts for the extra few minutes it takes (on average) for the Sun to return to the same orientation in the sky, as seen from a particular location on Earth.

However, Earth's orbit is not a perfect circle – in reality, it is slightly elliptical, so the actual distance to the Sun varies between 147.1 and 152.1 million km (91.4 and 94.5 million miles). As a result, the amount of heat and light Earth receives from the Sun changes slightly through the year (Earth is at its closest to the Sun in January and at its most distant in July).

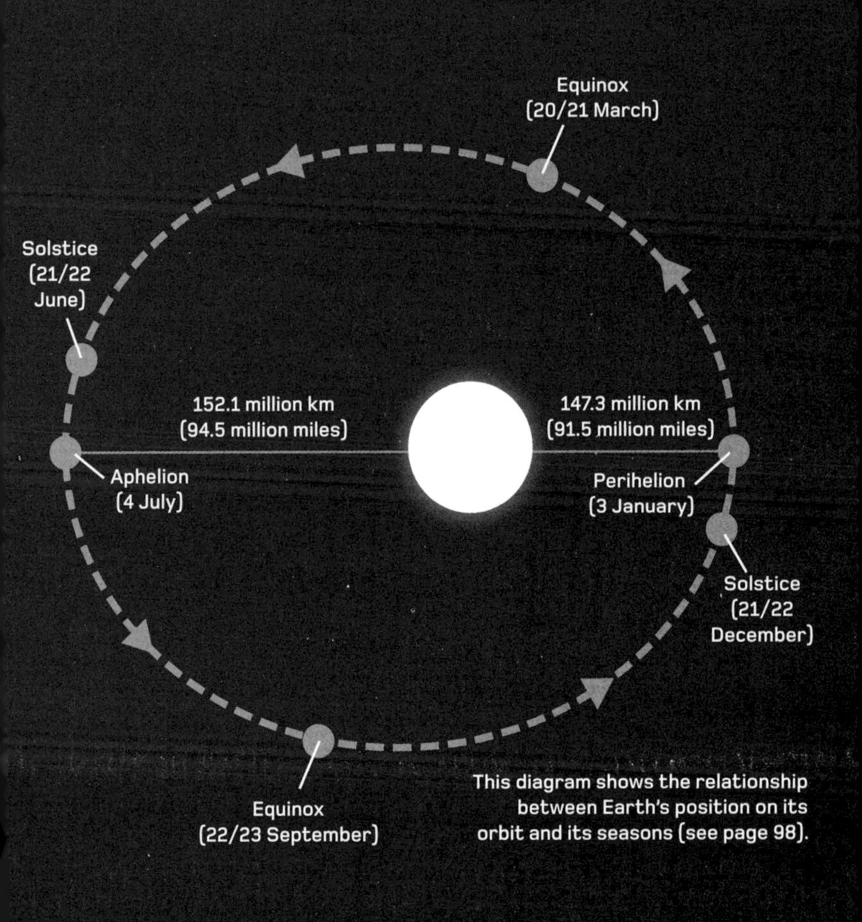

This diagram shows the relationship between Earth's position on its orbit and its seasons (see page 98).

The seasons

Like most other planets, Earth does not sit exactly 'upright' in its orbit. Instead, our planet's axis of rotation is tilted at an angle of 23.5 degrees. As Earth goes around the Sun during the course of a year, the axis remains pointing in the same absolute direction (roughly aligned with the direction of the famous 'North Star' Polaris). This means that for half the year, the northern hemisphere is, to some extent, angled towards the Sun, receiving the Sun's light and heat for more than half of each 24-hour day. For the other half of the year, however, that hemisphere is angled away from the Sun and receives considerably less sunlight. This is the origin of Earth's pattern of seasons – and, of course, when the northern hemisphere is tilted towards the Sun and experiencing summer, the southern hemisphere experiences winter and vice versa. The dates when Earth's axis points directly towards or away from the Sun are known as summer and winter solstices, while the two occasions in Earth's orbit where day and night are of equal length (and also the same in both hemispheres) are known as equinoxes.

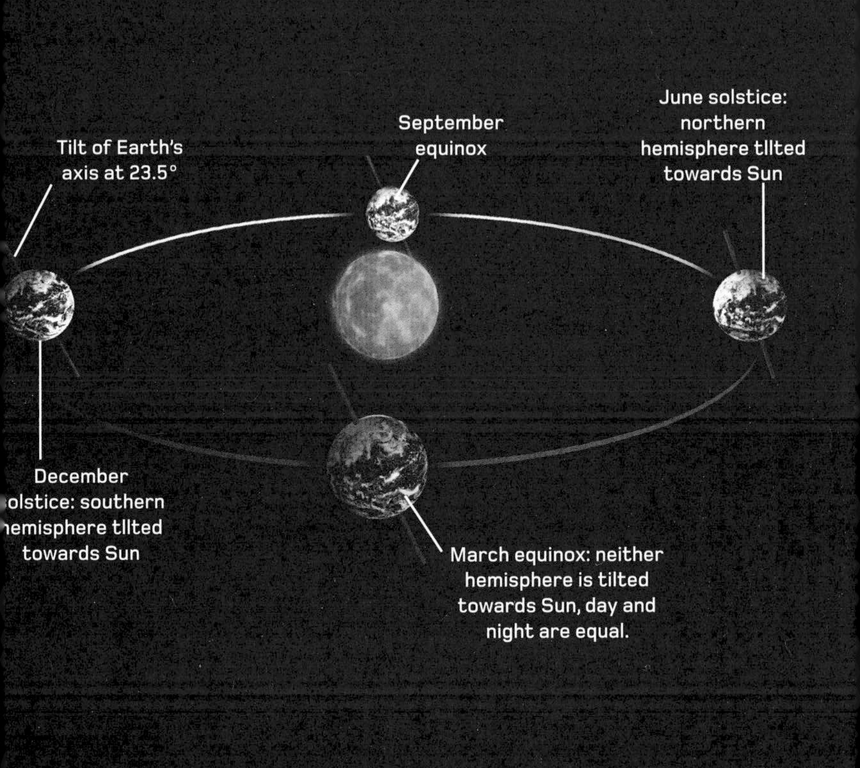
Tilt of Earth's axis at 23.5°
September equinox
June solstice: northern hemisphere tllted towards Sun
December
olstice: southern
emisphere tllted
towards Sun
March equinox: neither hemisphere is tilted towards Sun, day and night are equal.

Earth's interior

As the largest rocky planet, Earth naturally has the hottest interior. Trapped heat left over from Earth's formation is still slowly radiating away into space, and significant quantities of radioactive elements, such as uranium, release heat when they decay, helping to maintain the temperature.

Beneath the relatively thin crust (see page 102) and a lubricating layer called the asthenosphere, Earth's interior is mostly filled by its mantle – a deep layer of silicate rocks that are mobile over very long periods, slowly churning as hotter rocks push their way up from the regions around the core and cooler, denser ones slowly sink down. The transfer of heat through the mantle powers both surface volcanism and the slow motion of Earth's tectonic plates. Earth's core, meanwhile, occupies the central portion of the planet and seems to consist of two distinct regions – a liquid outer core of molten nickel and iron, and a solid inner core of the same elements that is slowly growing larger as the interior gradually cools.

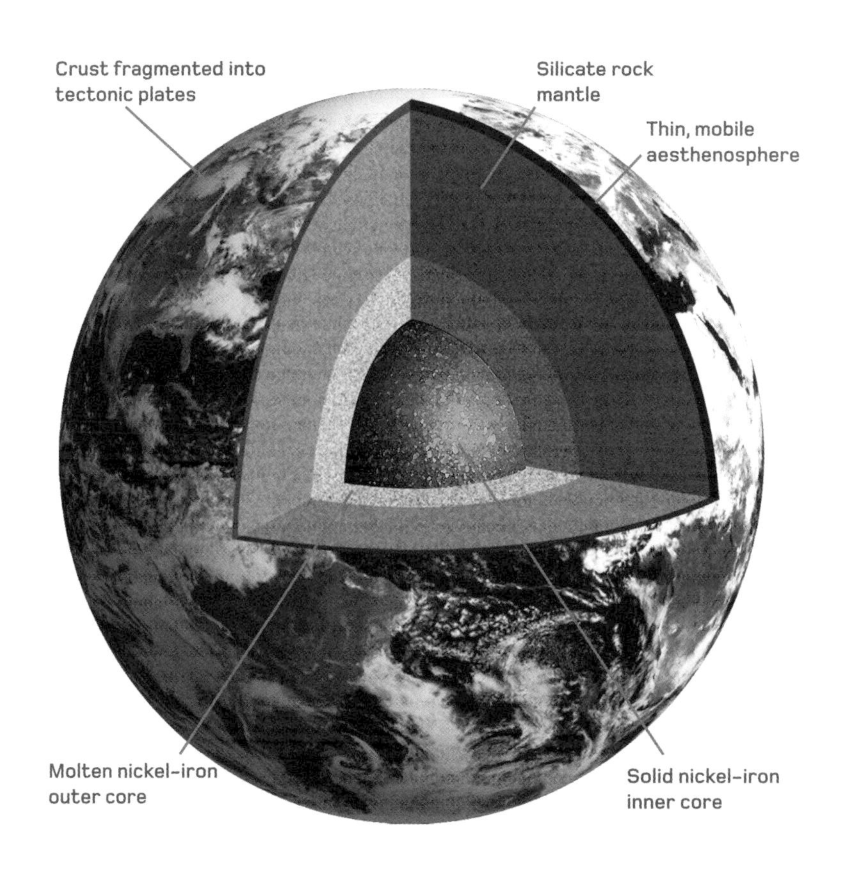
Crust fragmented into
tectonic plates
Silicate rock
mantle
Thin, mobile
aesthenosphere
Molten nickel-iron
outer core
Solid nickel-iron
inner core

Crustal geology

Earth's crust ranges in thickness from just a few kilometres beneath the ocean basins up to tens of kilometres deep beneath continents. Split into several dozen tectonic plates (eight major ones, ten minor ones and a host of 'microplates'), the crust effectively 'floats' on top of the aesthenosphere and underlying mantle, moving around at rates of millimetres per year. Plates may contain two distinct types of crust – thin, but dense, oceanic crust dominated by iron-rich 'mafic' rocks, such as basalt, and much thicker but lighter continental crust rich in aluminium-based 'felsic' rocks. Oceanic crust tends to be relatively young – it is created in areas where plates are separating (typically beneath the oceans), and destroyed where it is driven against continental rocks and forced downwards to melt in the mantle. Continental crust, in contrast, tends to be much older and is a complex mix of igneous rocks (formed from solidified molten lava), sedimentary rocks (formed from compression of fine particles themselves eroded out of other rocks) and metamorphic rocks (formed where other rock types are compressed and heated beneath the surface).

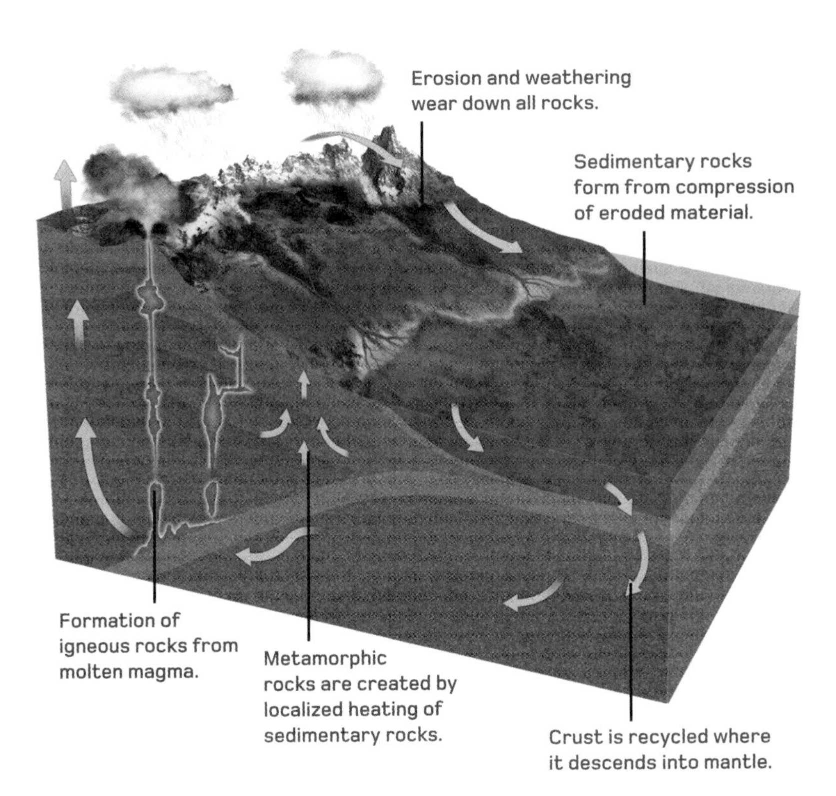

Erosion and weathering wear down all rocks.
Sedimentary rocks form from compression of eroded material.
Formation of igneous rocks from molten magma.
Metamorphic rocks are created by localized heating of sedimentary rocks.
Crust is recycled where it descends into mantle.

Earth's rock cycle

Atmosphere and oceans

Earth's atmospheric gases are dominated by nitrogen (78 per cent) and oxygen (21 per cent). Most of the remainder is inert argon, but a small amount of carbon dioxide traps heat close to the surface in a 'greenhouse effect' that is vital to keeping the planet warm, but rising dangerously due to human activity. Atmospheric circulation carries air from warm regions near the equator towards the colder poles, and is complicated by 'coriolis' forces created by Earth's rotation.

The oceans, meanwhile, account for 96.5 per cent of Earth's water, and are saline due to the presence of dissolved minerals. Evaporation from their surfaces pumps water vapour into the atmosphere, which condenses to form clouds and returns to the surface as fresh water precipitation (rain or snow), creating a water cycle that plays a key role in regulating the climate, sustaining life and shaping the landscape through erosion. Although 3.5 per cent of Earth's liquid water is fresh, most of this is locked into ice caps at the North and South Poles.

Life on Earth

Earth's abundant life makes it (so far as we know) unique in the solar system. Originating around 3.7 billion years ago, this life first took the form of simple 'prokaryote' microbes that reproduced themselves using instructions carried in the complex DNA molecule. Early microbes generated energy by processing or 'metabolizing' minerals and gases from the ancient atmosphere, gradually changing its composition over time. About 2.4 billion years ago, the first simple organisms to use the plantlike metabolic process of photosynthesis arose. The oxygen they produced cooled the planet, plunging it into a deep ice age, but opened the way for other metabolic processes. Around 1.6 billion years ago, the absorption of some of these innovative microbes into others allowed them to form symbiotic groups with specialized functions, known as 'eukaryotes'. It took about another billion years for the first multicellular life forms to arise, in which different eukaryotic cells perform specific functions. Since then, they have flourished, giving rise to the enormous complexity of today's plant, fungus and animal kingdoms.

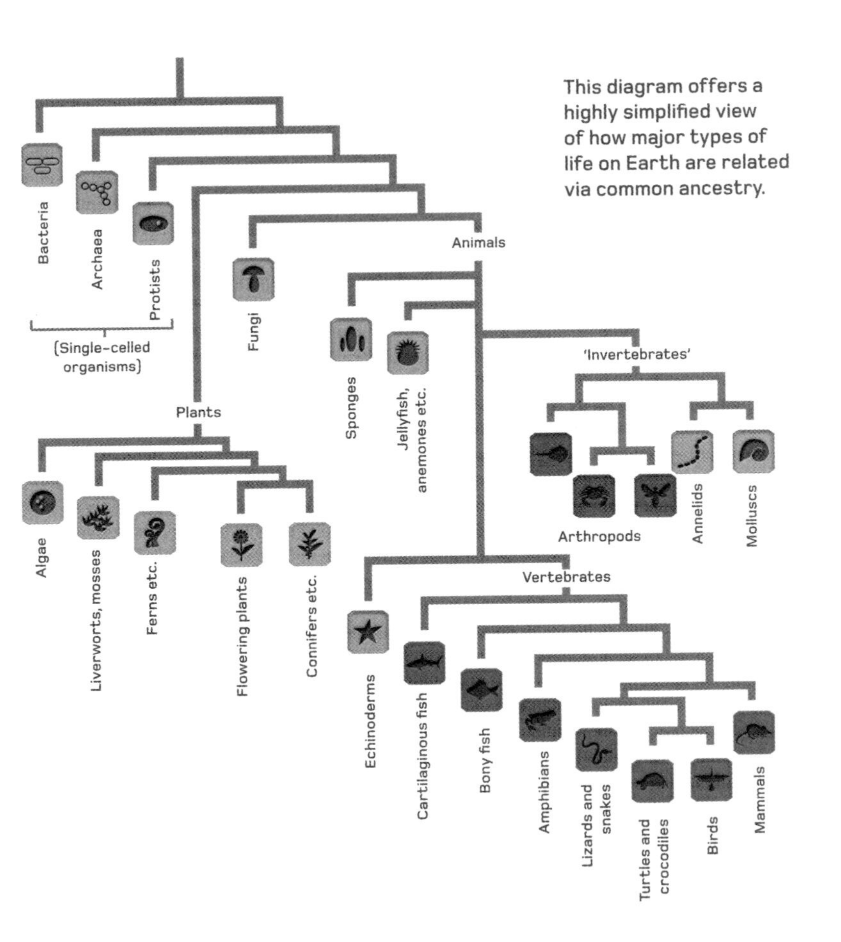
This diagram offers a highly simplified view of how major types of life on Earth are related via common ancestry.
Bacteria
Archaea
Protists
(Single-celled organisms)
Fungi
Animals
Plants
Sponges
Jellyfish, anemones etc.
'Invertebrates'
Arthropods
Annelids
Molluscs
Algae
Liverworts, mosses
Ferns etc.
Flowering plants
Connifers etc.
Vertebrates
Echinoderms
Cartilaginous fish
Bony fish
Amphibians
Lizards and snakes
Turtles and crocodiles
Birds
Mammals

Earth's climate

Our planet's climate is perhaps the most complex in the solar system, thanks to the intricate web of relationships between atmosphere, oceans and dry land. While the amount of heat we receive from the Sun varies slightly with solar activity, this has little effect on the overall climate (the average over longer periods of time), which is far more influenced by conditions on Earth. Overall, the climate system broadly conspires to maintain an environment suitable for water to move between solid, liquid and vapour forms in a 'water cycle' (see page 104). Several feedback mechanisms help keep the system in balance. One example is 'chemical weathering', in which falling rain takes on carbon dioxide from the atmosphere, and reacts with rocks to form carbonate minerals; a rise in carbon dioxide and a small increase in temperature should lead to increased rainfall and weathering that removes the gas from the atmosphere. However, such mechanisms typically work on timescales of thousands of years, and are less suited to coping with relatively sudden stresses such as human industrial emissions.

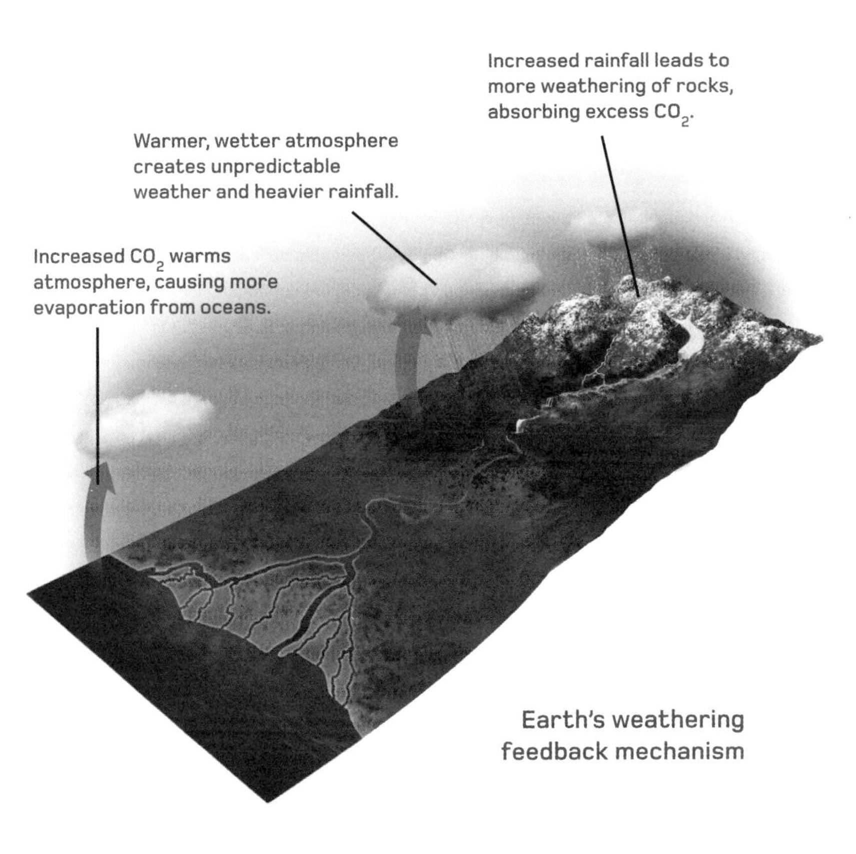
Increased rainfall leads to more weathering of rocks, absorbing excess CO_2.
Warmer, wetter atmosphere creates unpredictable weather and heavier rainfall.
Increased CO_2 warms atmosphere, causing more evaporation from oceans.
Earth's weathering feedback mechanism

The magnetosphere and aurorae

The space around our planet is dominated by a powerful magnetic field known as the magnetosphere. Its shape resembles the field around a familiar bar magnet, emerging from one magnetic pole, looping around the planet and re-entering at the other. The field originates in Earth's outer core, where it is generated by currents carried in the turbulent molten metal. Its orientation roughly matches Earth's axis of rotation, but the currents that create it periodically reverse, causing the field to flip its direction every few hundred thousand years.

The magnetosphere affects susceptible, electrically charged particles that pass through it, such as solar wind particles and high-energy 'cosmic rays' from the wider Universe. Most are deflected by the field, but others are swept up and energized to form doughnut-shaped radiation belts around the Earth. Particles channelled downwards around the magnetic poles collide with gases in the upper atmosphere to form the glowing 'aurorae', or northern and southern lights.

Earth's aurora australis (southern lights), photographed by an astronaut aboard the International Space Station

Meteors

As Earth moves along its orbit through space at an impressive 29.8 kilometres per second (66,600 miles per hour), it inevitably encounters other objects travelling at similar speeds. Most of this interplanetary matter is little more than dust or ice left behind in the wake of passing comets or asteroids. As they enter Earth's upper atmosphere, these tiny particles are heated by collisions with the sparse air molecules, burning up in short-lived streaks of light called 'meteors' or 'shooting stars'. The brightest, known as fireballs, can reach the lower atmosphere and outshine any star before they, too, are destroyed.

Most meteors enter the atmosphere from random directions, but some, still clinging to the orbits of their parent comets, form predictable meteor showers. They occur at the same time each year as Earth crosses their orbit and, thanks to perspective, they appear to radiate from a specific point in the sky. When Earth occasionally passes through a particularly dense region of such a comet trail, the result can be a spectacular meteor storm.

Meteorites

When a large or robust fragment of interplanetary rock enters Earth's atmosphere, it may survive its fiery journey and reach the ground partially intact, becoming a meteorite. About 500 such space rocks hit Earth each year – the last stages of their descent through the dense lower atmosphere are usually marked by a brilliant, slow-moving meteor called a 'bolide'. Meteorites have huge scientific value as samples of material from elsewhere in the solar system. Scientists divide them into several classes depending on how much geological 'processing' they have been through. Some are clearly fragments of core or mantle material from broken-up asteroids, while a few rare ones can even be traced to the Moon or Mars. The majority, however, known as 'chondrites', comprise tiny mineral spheres – raw material left unchanged from when the planets formed. Most are made from silicate minerals fused together by heat, but rare 'carbonaceous' chondrites seem to have avoided heating, allowing them to retain primordial water and delicate carbon-based chemicals from the ancient solar nebula.

The stony-iron Esquel meteorite is probably a fragment from an asteroid that began to develop a core and mantle before being shattered in an ancient collision.

Impacts from space

Although large impacts from space might seem rare in the brief span of human history, they are inevitable on geological timescales. If an object measuring tens of metres or larger strikes Earth's land surface, the energy unleashed can carve out a deep bowl-shaped crater many times bigger, spraying debris known as ejecta across an even wider area.

Earth's geological activity, atmosphere and abundant life can soften and disguise craters in a matter of decades, but the advent of satellite photography has revealed many hidden scars from ancient impacts. Some of the largest impacts have had a lasting effect on the history of life (see page 118), but even smaller ones can have potentially devastating consequences – the object that exploded in the air above Siberia in 1908 did not even leave a crater, but still carried enough energy to flatten a city-sized area of forest. For this reason, various projects are underway to map the distribution of potentially threatening objects in Earth-crossing orbits.

Arizona's famous 'Meteor Crater', some 1,200 metres (3,900 ft) across, was formed by the impact of a 50-metre (160-ft) meteorite about 50,000 years ago.

Chicxulub Basin

Earth's most significant surviving impact crater has long since been buried beneath the surface by geological activity. The 180-km-wide (112 miles) Chicxulub crater lies hidden beneath the modern Gulf of Mexico and was only discovered by accident during petroleum exploration in the 1980s. Chicxulub's huge scale suggests that it was formed by an asteroid or comet some 10–15 km (6–9 miles) across, and geological clues suggest that the impact sent clouds of pulverized rock high into the atmosphere, from where it slowly settled across the planet a little less than 66 million years ago.

Chicxulub owes its fame to a key event in the history of life on Earth – it coincides precisely with the mass extinction that saw the disappearance of the large reptilian dinosaurs and countless other species, paving the way for the rise of mammals. Although the precise chain of events is still disputed, it seems that widespread climate cooling played a key role, triggered by dust flung into the atmosphere during the impact.

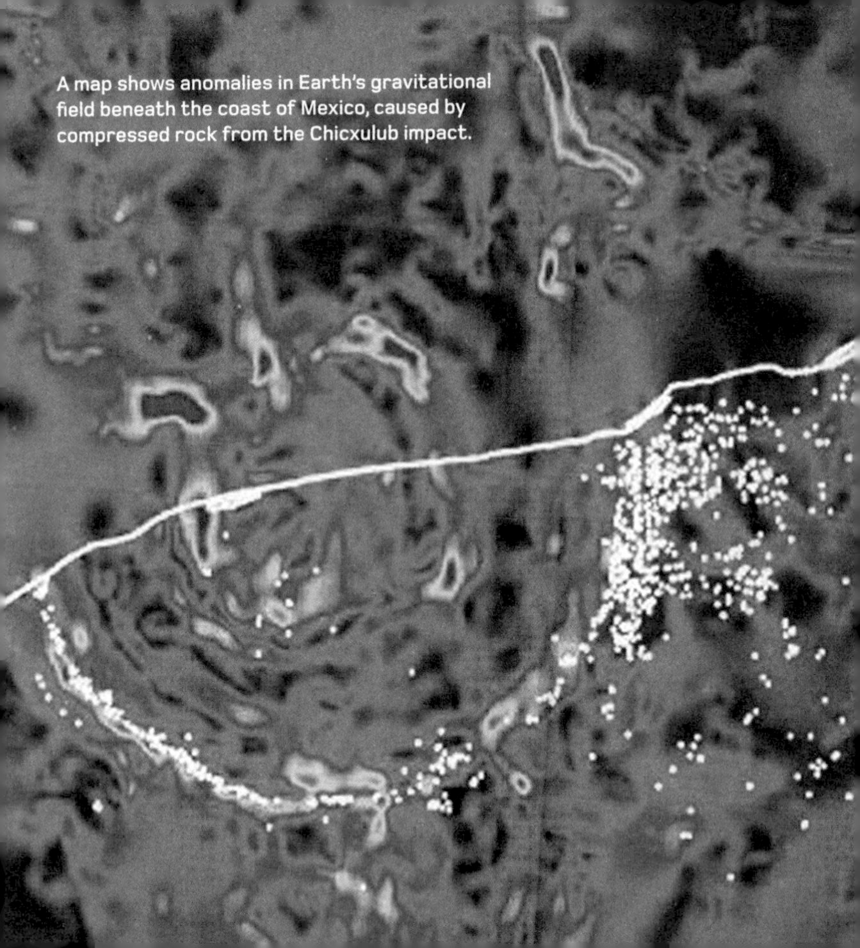

A map shows anomalies in Earth's gravitational field beneath the coast of Mexico, caused by compressed rock from the Chicxulub impact.

Earth's Moon

Our planet's Moon (typically given a capital 'M' to distinguish it from the satellites of other planets) is the most prominent object in the night sky. Orbiting Earth in 27.3 days at an average distance of 384,400 km (238,850 miles), tidal forces (see page 122) have long since slowed its rotation to the same period, so that one hemisphere is permanently turned towards Earth. The amount of the Earth-facing side illuminated by sunlight, however, is continually changing, giving rise to a familiar cycle of phases that repeats every 29.5 days (as the Moon comes back into the same alignment with the Sun). With a diameter of 3,474 km (2,158 miles), the Moon is just over one- quarter the size of Earth, and has 1.2 per cent of its mass. This makes it by far the solar system's largest satellite relative to the size of a major planet, and gives it considerable influence over Earth. However, lunar gravity (at about one-sixth the strength of Earth's) is far too weak to hold onto an atmosphere; as a result, the Moon is an airless ball of rock on which temperatures range between –170°C and 220°C (–274°F and 428°F).

The Earth–Moon system

Earth and Moon are so similar in size and exert so much influence on each other that some astronomers have argued they should really be considered as a 'double planet'. The most obvious effect is in the form of tides – forces caused by 'fall-off' in the strength of each world's gravity as felt between the near and far sides of the other body. On Earth, tides create a 'bulge' several metres high in the oceans lying closest to the Moon, and a slightly smaller bulge in oceans on the opposite side of the planet. The bulges stay in roughly the same position as Earth rotates beneath them – hence, there are two high tides per day. The situation is further complicated by weaker tidal bulges caused by the Sun. Tides experienced by the Moon were once much stronger, though less obvious because of its solid rocky composition. Over billions of years, forces acting on the Moon's small but noticeable tidal bulges were reduced as its rotation slowed to match its orbital period. Today, the Moon suffers small tides as its distance from Earth varies around its monthly orbit, giving rise to occasional 'moonquakes'.

Seas and highlands

The Moon's surface is divided into two main types of terrain – bright, heavily cratered highlands and darker, smoother 'seas' or 'maria'. Satellite images and robot landers have shown that the rough highland terrain is saturated with craters down to microscopic scales, and which date back to the earliest lunar history. With no water or wind and little geological activity, the main factor eroding ancient craters is the arrival of later impacts on top of them. The lunar maria, in contrast, are vast plains of solidified lava, formed between 3 and 4 billion years ago, in a period when low-lying regions of the Moon were vulnerable to volcanic eruptions that flooded the deepest impact basins. This activity obliterated heavy cratering from the period of the Late Heavy Bombardment (see page 36) and left a clean slate onto which more recent craters have accumulated at a much-reduced rate. Nevertheless, bombardments have pounded the upper few metres of their surfaces into a jumble of rocks and dust of different sizes, known as 'regolith'.

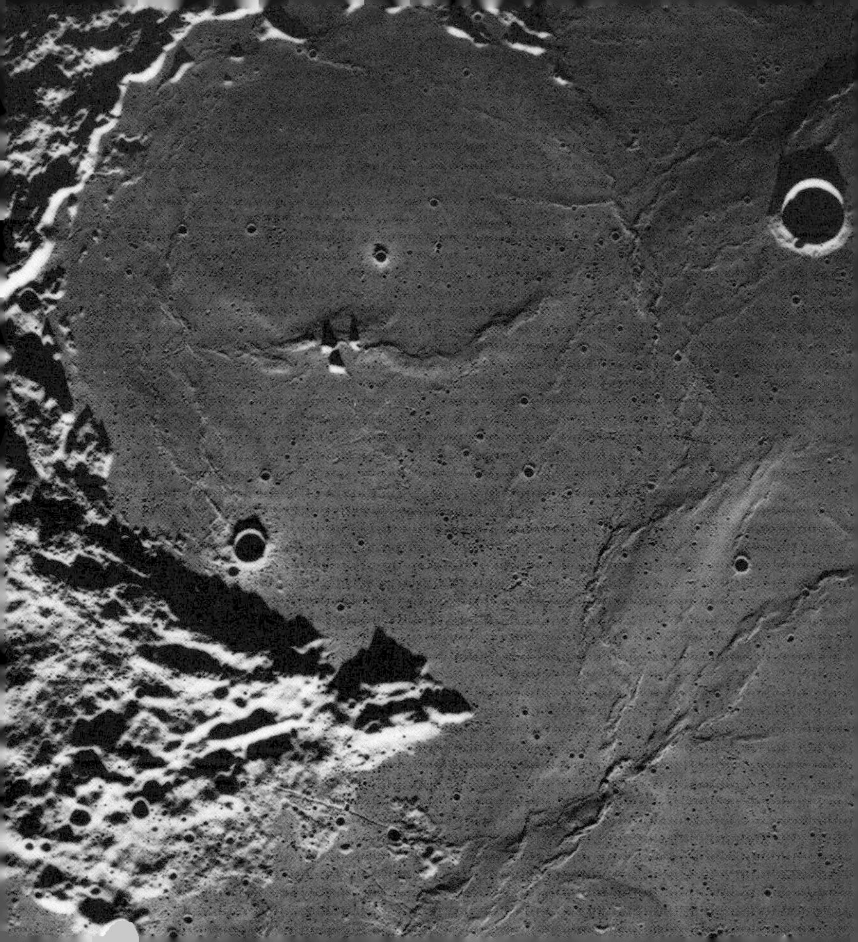

Origin of the Moon

The Moon's origin has been a subject of fierce debate for almost two centuries; astronomers have found it hard to explain why such a large satellite would form around a relatively small planet. Early theories, such as the ejection of the Moon from a fast-spinning ancient Earth, contradicted basic laws of physics, while the once-popular idea of the Moon as a small rogue planet captured from elsewhere in the solar system has faltered in the face of evidence that the rocks of Earth and Moon are almost identical. The most credible modern theory, known as the 'Big Splash' hypothesis, appears to solve most of these problems. It proposes that, shortly after its formation, Earth was involved in a collision with a rogue Mars-sized planet (sometimes called Theia). The collision saw Earth and Theia merge together, but in the process ejected vast amounts of debris from both worlds into orbit, where it rapidly coalesced to form the Moon. However, the theory is not without its problems – particularly when it comes to delivering a satellite with the Moon's own unique geology.

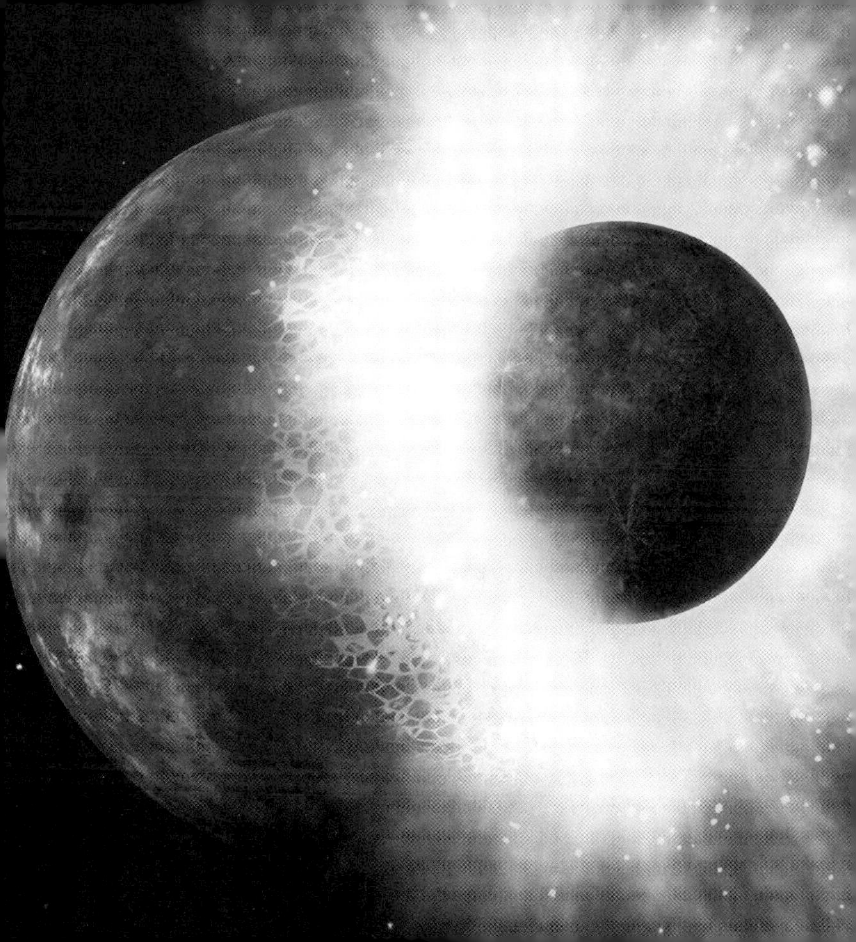

Mare Crisium

One of the most distinctive features of the lunar nearside, the Mare Crisium is a relatively small 'sea' some 555 km (345 miles) in diameter. Lying in the Moon's northeast quadrant, its appearance is significantly foreshortened by the Moon's curvature, making it look oval, rather than circular in shape. The sea's surface is mostly flat, but 'wrinkle ridges' around the edges show how solidifying lava once piled up as it lapped against the surrounding highland shores. The impact basin in which the mare lies is about 3.9 billion years old, and analysis of rock samples returned by the unmanned Soviet Luna 24 probe in the 1970s suggests that the basaltic lavas within it formed in three episodes between about 3.5 and 2.5 billion years ago.

Like several other lunar seas, the Mare Crisium coincides with an area of unusually high density known as a mascon (mass concentration). Thanks to their higher gravity, these poorly understood phenomena present a significant hazard to satellites and spacecraft orbiting the Moon.

Montes Apenninus

The Moon's most notable mountain range, the Montes Apenninus or Lunar Apennines, forms a curving chain that stretches across some 600 km (370 miles) in mid-northern latitudes of the lunar nearside. Rather than being the result of uplift from Earth-like tectonic forces, the range is simply the broken-down rim of the huge Imbrium impact basin, formed about 3.9 billion years ago and subsequently disguised by eruptions of mare basalt lavas that flowed between its peaks.

The range's highest point, Mons Huygens, lies towards the range's southern end and rises some 5.5 km (3.4 miles) above the nearby lava plains. At the northeast end, meanwhile, stands Mons Hadley Delta, a 3.6-km (2.2-mile) triangular peak close to the 1971 landing site of the Apollo 15 mission. Another major feature of this region is Hadley Rille, a serpentine, round-bottomed valley some 135 km (84 miles) long and up to a kilometre (0.6 miles) across, formed by the collapse of an abandoned subterranean lava channel.

Tycho crater

One of the Moon's most prominent craters, Tycho lies in the southern highlands and draws attention to itself at the centre of a system of bright 'rays' (material sprayed out during formation that travelled for huge distances in the weak gravity before settling back to the surface) that spread for up to 1,500 km (930 miles) across the lunar surface. Tycho presents a particularly perfect example of a lunar crater, with a diameter of 85 km (53 miles) and a depth of 4.8 km (3 miles). Terraced walls and a distinctive central mountain were both created in the last stages of the crater's formation, when the rim of an initially bowl-like depression became too steep and slumped downwards, pushing up the middle of the crater floor in the process.

Tycho is one of the youngest large craters on the Moon – it has remained in pristine condition for some 108 million years, as have its surrounding rays. Tycho is also surrounded by a series of smaller satellite craters, formed as more substantial chunks of ejecta fell back closer to the impact site.

The lunar far side

Unseen by astronomers until the Soviet Luna 3 probe sent back the first grainy images in October 1959, the far side of the Moon appears distinctly different from the near side, thanks to an almost complete lack of dark lunar seas. The 180-km (112-mile) crater Tsiolkovskiy is one of the few obvious dark features, while the huge South Pole-Aitken Basin, some 2,500 km (1,550 miles) wide and 13 km (8 miles) deep, remains entirely unfilled.

Scientists believe the difference in general appearance between near and far sides is due to the same tidal forces that long ago slowed the Moon's rotation to match its orbital period. Earth's gravity, which tugs more strongly on the nearer side of the Moon, seems to have pulled the Moon's molten core a few kilometres closer to its Earth-facing side, causing greater volcanic activity on that side. Quite why Tsiolkovskiy should be an exception to this general pattern remains unclear.

Mars

The outermost of the solar system's rocky planets, Mars is also the second smallest, with a diameter of 6,779 km (4,212 miles) – just over half that of Earth. The planet's 687-day orbit is markedly elliptical, ranging between 1.38 and 1.67 AU from the Sun, but in other respects Mars is remarkably Earth-like; it spins on its axis in 24 hours and 37 minutes and has an axial tilt of 25.19 degrees that produces a familiar pattern of seasons.

The Martian atmosphere, however, is very different from Earth's – composed almost entirely of carbon dioxide, it is so thin that it exerts just one per cent of Earth's atmospheric pressure. Nevertheless, it acts as a surprisingly effective insulator, so while temperatures can fall as low as –143°C (–225°F) at the poles in winter, they can also rise as high as 35°C (95°F) around the equator in summer. When Mars is at its closest to the Sun during the southern summer, heat in the atmosphere can lift huge amounts of fine red dust into the air, engulfing the entire planet in a global dust storm that may take months to subside.

Interior of Mars

As with the other rocky planets, heavier elements inside Mars have sunk towards the centre while lighter ones have risen to the surface, forming a layered structure of core, mantle and crust. Estimates based on the Martian gravitational field suggest that the planet's core is lighter than those of the other inner planets, probably due to a significant amount of sulfur mixed with its iron and nickel. The lack of a strong magnetic field around Mars today suggests that the core no longer contains liquid metal. However, the presence of sulfur could have lowered its freezing point, allowing the core to remain molten for longer and generating a field for much of the planet's history that would have helped to protect a once-thicker atmosphere from the stripping effect of the solar wind. Heat rising through the silicate minerals of the overlying mantle powered volcanic activity for much of Martian history. A fair amount of this rising heat seems to have concentrated in a single column, creating a 'hot spot' beneath a region called Tharsis, giving rise to huge volcanoes and a vast bulge of accumulated lava flows – the 'Tharsis Rise' – that stands up to 7 km (4.4 miles) above its surroundings.

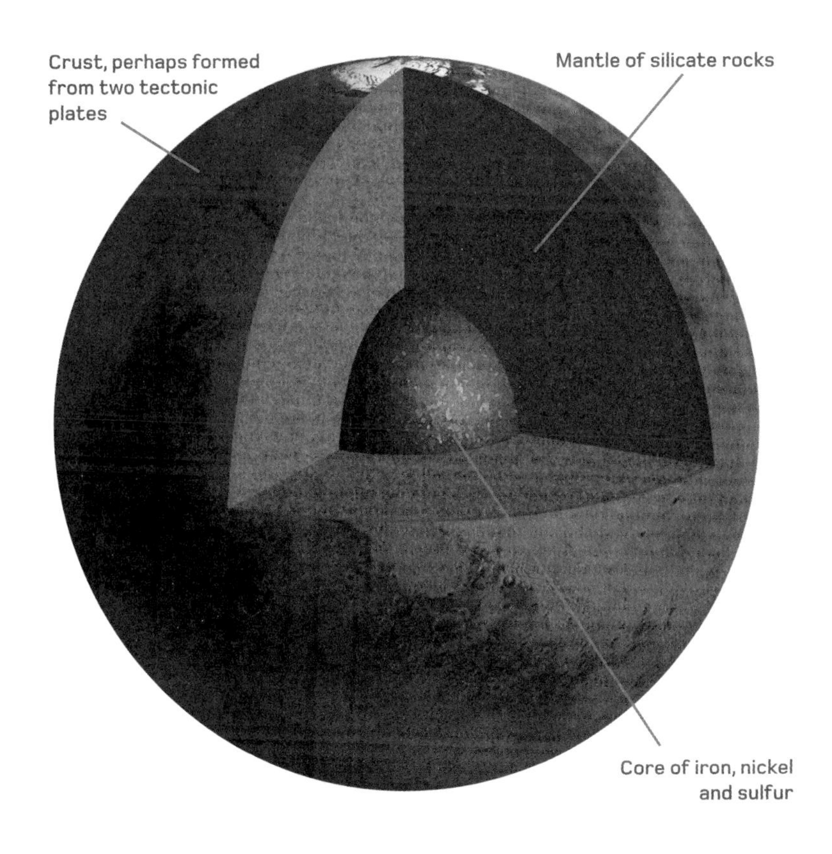
Crust, perhaps formed from two tectonic plates
Mantle of silicate rocks
Core of iron, nickel and sulfur

Martian geography

The Martian landscape can be divided into two distinct terrains – relatively smooth lowlands in the north (with the huge bulge of the Tharsis Rise emerging from them), and heavily cratered southern highlands. Seen from space, two of the most prominent features are Syrtis Major, a dark, wedge-shaped plain just north of the equator, and Hellas Planitia, an ancient dust-filled impact basin in the mid-southern latitudes. The boundary between highlands and plains is marked in many places by 'fretted terrain', a chaotic jumble of elevated flat-topped mesas with valleys running between them.

The origins of this 'Martian surface dichotomy' are much debated, but it's clear that some major event either erased the record of early cratering across the planet's northern hemisphere, or prevented this hemisphere's crust from solidifying at all until the heaviest cratering had subsided. Possible explanations include one or more major impacts from space, or large-scale tectonic processes in early Martian history.

Olympus Mons

The largest volcano on Mars, Olympus Mons is also the biggest mountain in the solar system, with a shallow dome shape more than 620 km (385 miles) across, rising to 21.3 km (13.2 miles) above the average Martian surface level. It lies just off the northwestern side of the Tharsis Rise, and is an enormous shield volcano, created by lava that erupted from multiple fissures along its flanks. Like Earth's Hawaiian island volcano chain, Olympus Mons sits above a 'hot spot' in the underlying mantle. However, because the red planet lacks plate tectonics, the eruptions powered by this source have remained in one place, building layer upon layer of lava over an estimated three billion years. The 85-km (53-mile) crater complex at the peak is a 'caldera' – a series of pits created by subsidence when the supporting pressure of molten magma in the underlying mantle was withdrawn at various times. Although the volcano appears dormant today, some of its youngest lava flows have been dated to as recently as two million years ago, so Olympus Mons may not yet be entirely extinct.

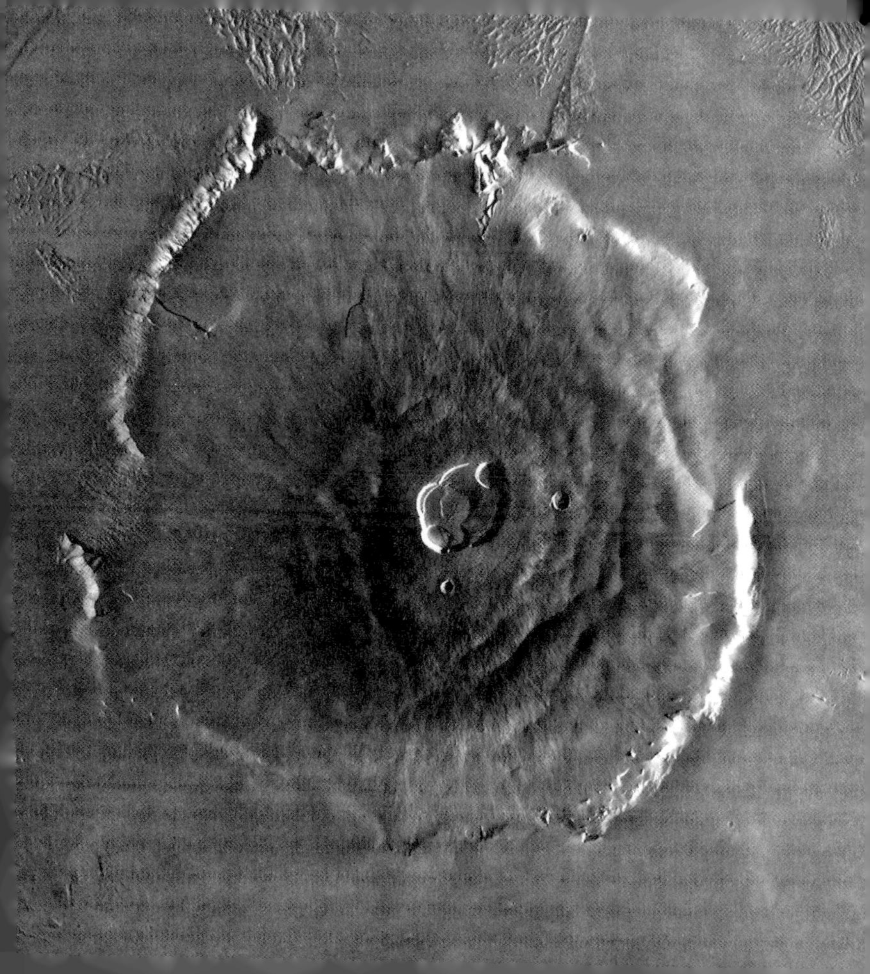

Valles Marineris

One of the most spectacular features in the solar system, the Valles Marineris (Mariner Valleys, named after the Mariner 9 space probe that discovered them in 1972) form a continent-sized scar across the Martian surface. More than 4,000 km (2,500 miles) long, and up to 10 km (6 miles) deep in places, they dwarf Earth's own Grand Canyon.

Unlike the Grand Canyon, however, this valley system is not the result of water erosion. Running along the southern edge of the volcanic Tharsis Rise, it is thought to have been cracked open by stresses in the Martian crust due to the enormous weight of accumulated lavas. This is not to say that the valleys have always been entirely dry, however – as the lowest-lying part of the surface, its higher air pressure would have made this a favourable location for surface water in the distant past. Confirming this, analysis of minerals exposed by landslips in the vast central depression known as Melas Chasma suggests that the area was once covered by a large lake.

This 3D view shows the broad 'Candor Chasma' portion of the Valles Marineris complex.

Chryse Planitia

Lying to the west of the Tharsis Rise and north of the Martian equator, Chryse Planitia is a low-lying circular plain covered in broad channels that wind between teardrop-shaped mounds. Occupying an ancient impact basin that formed early in Martian history, the overall terrain is reminiscent of the 'channelled scablands' found on Earth in the northwestern United States and other areas. In our own planet's case, such features are linked to the melting of glaciers at the end of the last ice age – vast lakes, held back for centuries by walls of ice, were suddenly released to cause catastrophic floods that scoured the landscape. On Mars, it seems something similar happened – billions of years ago, a large canyon system called Kasei Valles, which empties into its western side, formed one of the main outflows for water draining off the vast Tharsis Rise. In 1997, NASA's Mars Pathfinder lander arrived in Ares Vallis, a similar outflow channel that runs into the southeast edge of Chryse, where its Sojourner rover investigated a variety of highland rocks deposited by ancient floods.

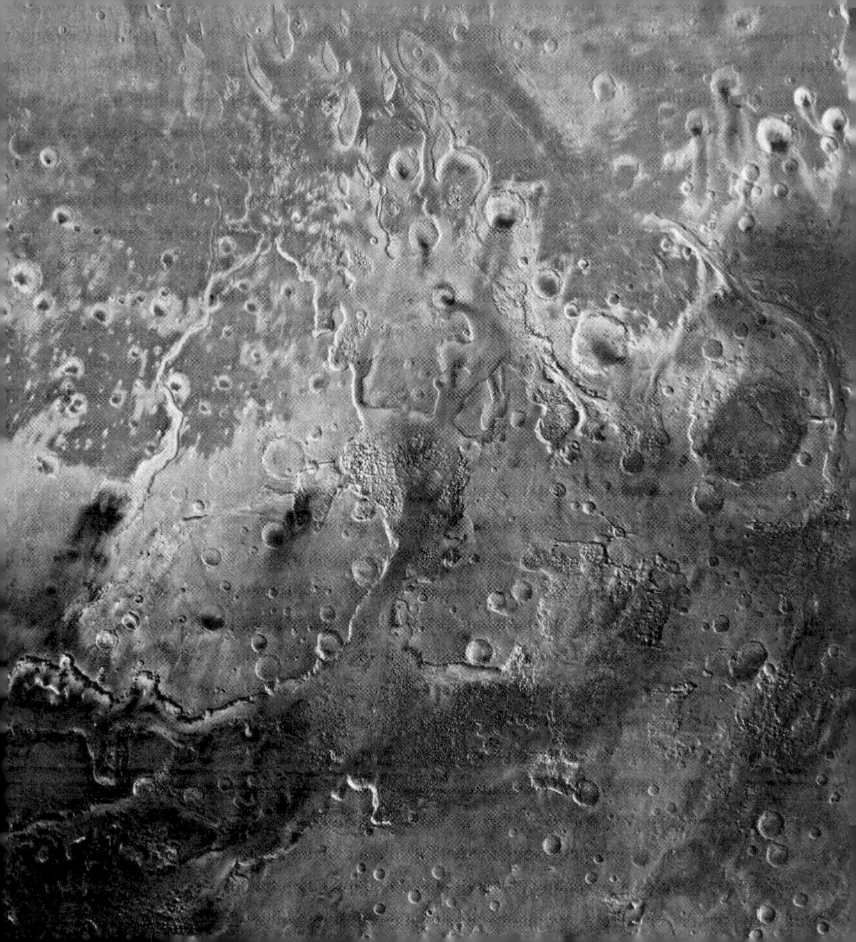

Polar caps

Seen from Earth, the most prominent Martian features are two distinctive ice caps that grow and shrink with the changing seasons. Each cap consists of a permanent body of deep-frozen water overlaid with a changeable layer of frozen carbon dioxide (CO_2 or 'dry ice'). About 25 per cent of all the atmosphere's CO_2 settles as frost or snow during a particular hemisphere's autumn, and returns to the air by a process called sublimation (transition directly from solid to gas) in spring. This activity drives prevailing winds.

As small amounts of ice mixed with the ubiquitous Martian dust are left behind in each successive cycle, the polar caps have built up a complex layered structure kilometres deep. Spiral ravines that cut into the ice caps show where prevailing winds carry the sublimated carbon dioxide away. However, water ice extends far beyond the visible limits of the polar caps. Mixed with the soil it forms permafrost beneath the northern plains, while in purer form it produces slow-flowing glacial features, disguised by a thin coating of red dust, amid the craters of the southern highlands.

A perspective view of the Martian north polar ice cap

The wet Martian past

Evidence that Mars had water flowing on its surface in its ancient past is abundant. Features that resemble the work of water on Earth, including winding river valleys and catastrophic flood features, such as Chryse (see page 146), were discovered in satellite images during the 1970s. However, alternative explanations remained on the table until the 2000s, when a new wave of orbiters and landers confirmed the widespread presence of hydrated minerals formed in the presence of liquid water.

At first, many scientists believed that the Martian water had been lost forever – evaporated into the atmosphere by ancient climate change and then blown away into space on the solar wind. However, the discovery that much of the Martian soil is ice-rich permafrost suggests a different story. It's now recognized that the red planet goes through long-term climate cycles thanks to slow variations in its orbit. Mars may even now be emerging from a long ice age, and could develop a thicker atmosphere and surface water once again in the future.

Artist's impression of the shoreline of an ancient Martian sea

Present-day water

Evidence for liquid water running on the surface of Mars today (or concealed in underground watercourses) would be a huge scientific breakthrough, but despite several significant discoveries, the clinching proof remains frustratingly elusive. During the 1990s, images sent back from NASA's Mars Global Surveyor satellite revealed geologically recent gullies along the walls of several Martian valleys, apparently created by liquid seeping out from a layer just beneath the surface. However, while such features on Earth would immediately be interpreted as evidence of water, alternative explanations, perhaps involving the escape of frozen carbon dioxide, cannot be ruled out in the case of Mars. In 2011, spectacular pictures from the Mars Reconnaissance Orbiter showed dark streaks on the slopes of a crater called Newton, changing with the Martian seasons and apparently running downhill. These, too, have been explained in terms of subsurface water but, frustratingly, attempts to detect the water directly (using satellite spectrographs) have so far proved inconclusive.

The rim of Newton crater shows seasonal dark streaks, perhaps linked to water, running down the crater wall at bottom.

Life on Mars?

Although the idea of plants and advanced life forms on Mars was sadly abandoned as the planet's true barren nature became clear in the 1960s and 1970s, new discoveries about life in extreme environments, and conditions on Mars itself, have kept open the possibility of simpler past or present life. Claims of 'microfossils' and biochemicals in a sample of Mars rock that crashed to Earth in Antarctica first grabbed headlines around the world in 1996. However, alternative explanations have since been put forward and, though the discoverers stand by their findings, the debate seems unlikely to be settled without new evidence. So far, the only probes to search directly for signs of life were NASA's 1970s Viking landers (see page 384), although their results were frustratingly inconclusive. More recently, scientists have discovered intriguing seasonal emissions of methane – a rapidly degrading chemical generated by only a few processes, including volcanism and life. In 2018, NASA's Curiosity rover (see page 394) discovered traces of carbon-based organic chemicals in the soil – not conclusive proof of life itself, but potential traces of its presence.

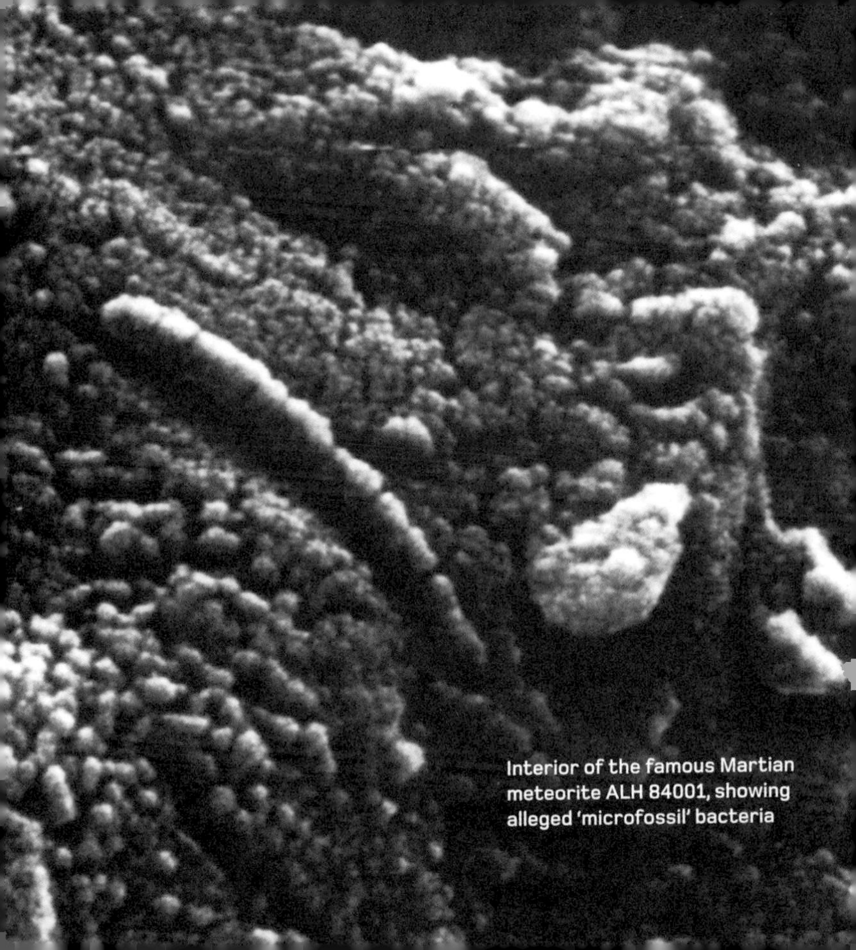

Interior of the famous Martian meteorite ALH 84001, showing alleged 'microfossil' bacteria

Phobos

Mars has two small moons, named after the sons of the warrior god Ares in ancient Greek mythology. Phobos is the larger of the two – a potato-shaped lump of rock with dimensions of 27 × 22 × 18 km (17 × 14 × 11 miles). Its surface is dominated by a large crater called Stickney, from which curious parallel furrows radiate across the surface. One possibility is that these scrapes were made by ejecta that tumbled across the surface in the weak gravity following the initial Stickney impact, but later escaped into space.

Phobos is also the closer of the two satellites to Mars. It orbits at an altitude of just 6,000 km (3,700 miles) above the surface (60 times closer to Mars than the Moon is to Earth), and circles the planet in 7 hours 39 minutes, clipping the thin upper atmosphere as it does so. In around 40 million years, this orbit will decay to instability and Phobos will begin to tumble out of control, probably breaking up before its fragments crash into Mars itself, forming a new family of craters.

Deimos

The smaller and more distant of the two Martian moons, Deimos orbits its parent planet in a relatively sedate 30 hours 18 minutes. In contrast to the pockmarked Phobos, its surface is relatively smooth, suggesting that impact craters on its surface have been filled and softened by large amounts of dust.

Astronomers used to assume that these two small Martian moons must be asteroids captured into Martian orbit, although their peculiarly unique surface chemistry and reddish colour led to the suggestion that they might be stray Trojans from around the orbit of Jupiter (see page 192). Recent computer models, however, suggest that a precise capture of even one asteroid by the weak Martian gravity, let alone two, would require a very precise set of circumstances that are unlikely to have occurred in the entire history of the solar system. An increasingly attractive idea, therefore, is that Phobos and Deimos formed in a similar way to Earth's Moon, coalescing out of debris ejected after a major impact on the parent planet's surface.

The asteroid belt

Although the inner solar system is scattered with asteroids, the vast majority lie within a broad, doughnut-shaped belt between the orbits of Mars and Jupiter. With an inner edge around 2.1 AU and an outer cut-off 3.3 AU from the Sun, the belt contains several hundred million objects of varying size, scattered in mostly empty space (though collisions and close encounters are relatively frequent on astronomical timescales). The first objects to be found here were discovered in the early 19th century, as astronomers looked for traces of a planet in the apparent 'gap' between the orbits of Mars and Jupiter.

Only around 200 asteroids have diameters greater than 100 km (60 miles) and the entire belt's combined mass amounts to just four per cent of Earth's Moon. Nevertheless, the asteroid belt is commonly held to mark a region where Jupiter's powerful gravity prevented the formation of a more substantial fifth rocky planet – most of its potential material was either flung towards the Sun or completely ejected from the inner solar system.

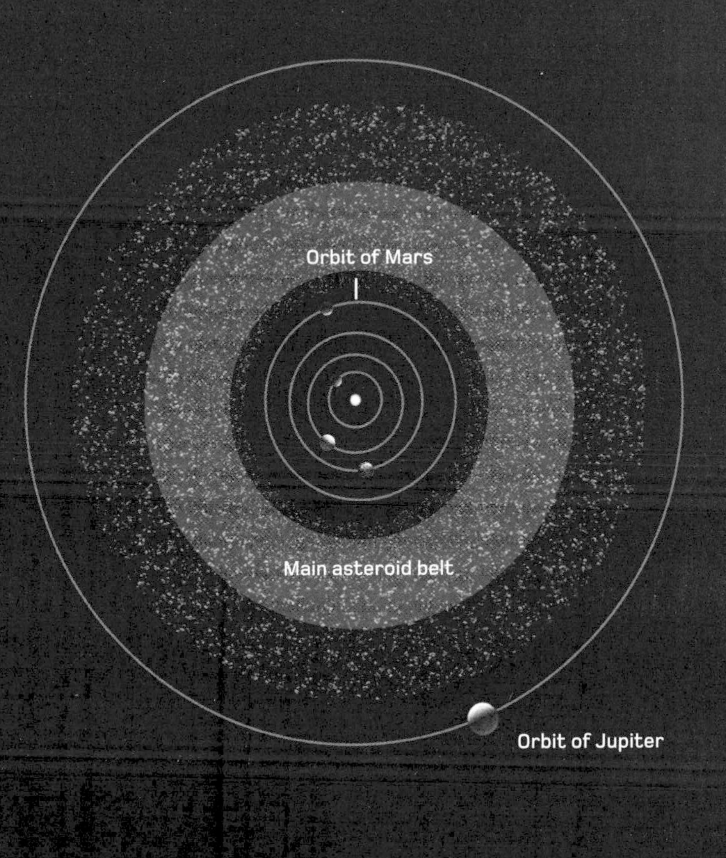
Orbit of Mars
Main asteroid belt
Orbit of Jupiter

Kirkwood gaps

By 1866, enough asteroids had been discovered for US astronomer Daniel Kirkwood to note the existence of a number of 'gaps' in the asteroid belt. These asteroid-free regions exist because the orbital periods of any object orbiting here would bring it into repeated close alignment with Jupiter. Therefore, any asteroids that fall by chance into such 'resonant' orbits are soon kicked out onto a different path. In 1898, the first refugee from these regions, a so-called Near Earth Asteroid catalogued as 433 Eros (see pages 164 and 190), was discovered by German astronomer Gustav Witt.

The major Kirkwood gaps are located around 2.5, 2.82 and 2.92 AU from the Sun, in regions where orbital periods are one-third, two-fifths and three-sevenths of Jupiter's, respectively. Two further 'gaps', at 2.06 and 3.27 AU, are generally taken as the inner and outer edges of the main asteroid belt itself. Curiously, some other simple resonances seem to concentrate asteroids rather than ejecting them, producing asteroid 'families'.

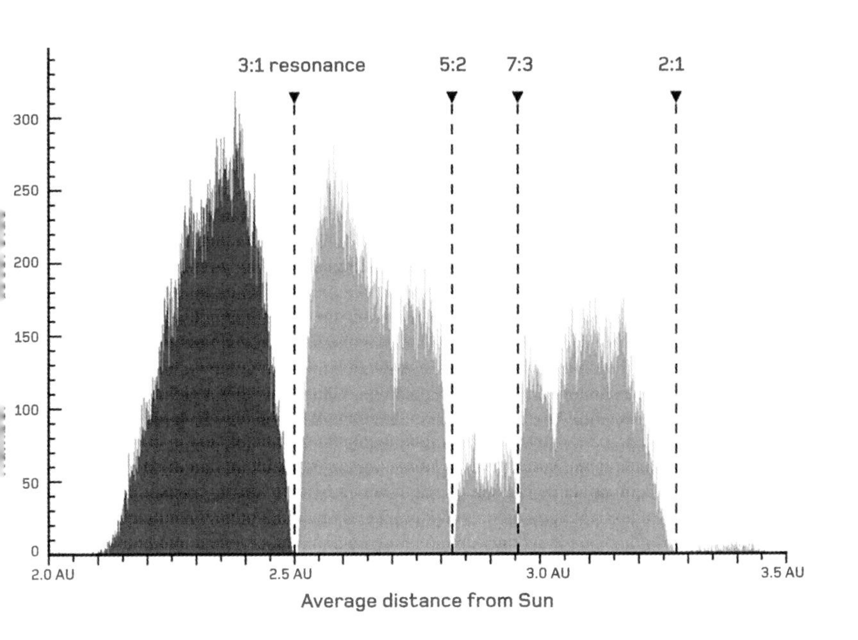

Kirkwood gaps and orbital resonances with Jupiter
3:1 resonance
5:2
7:3
2:1
300
250
200
150
100
50
0
2.0 AU
2.5 AU
3.0 AU
3.5 AU
Average distance from Sun

Near-Earth Objects

The discovery of asteroids orbiting within the main belt, on elliptical paths that bring them close to the strong gravity of the rocky planets, seems surprising at first. Any object in such an orbit is living on borrowed time and will inevitably experience close planetary encounters that will alter its orbit, perhaps ejecting it from the solar system, hurling it into the Sun, or ending with a spectacular planetary collision.

The most likely explanation for the significant number of such bodies, then, is that they are recent arrivals from elsewhere in the solar system. Most probably originated in the main asteroid belt and fell into their present orbits after straying into a resonant orbit with Jupiter (see page 162). Known as either Near-Earth Asteroids or, more inclusively, Near-Earth Objects (NEOs), their orbits are typically unstable over timescales of millions of years, so while we might predict their paths and the likelihood of close approaches to Earth in the short term, the more distant future is far harder to determine.

Types of NEOs

Near-Earth Objects (NEOs) are divided into several different groups depending on the shape of their orbits. Amor asteroids orbit further from the Sun than Earth, and while they come close to our planet they do not cross its orbit. Atiras or Apopheles, in contrast, have orbits closer to the Sun than Earth. The Aten and Apollo groups, however, are 'Earth-crossers'. Aten asteroids spend most of their time at less than 1 AU from the Sun, but some of it beyond Earth's orbit, while the reverse is true for Apollos. Not all of these asteroids present a risk to Earth, of course – some (like Cruithne, see page 187) follow orbits that keep them away from our planet, while others have inclined orbits that do not intersect with Earth's. 1862 Apollo itself, after which this category is named, is a roughly 1.5-km (0.9-mile) asteroid first spotted in 1932, but subsequently lost for more than 40 years. It follows a highly elliptical trajectory and crosses the orbits of Venus and Mars as well as that of Earth – the potential gravitational influence of these three bodies makes its path hard to predict beyond a few hundred years in the future.

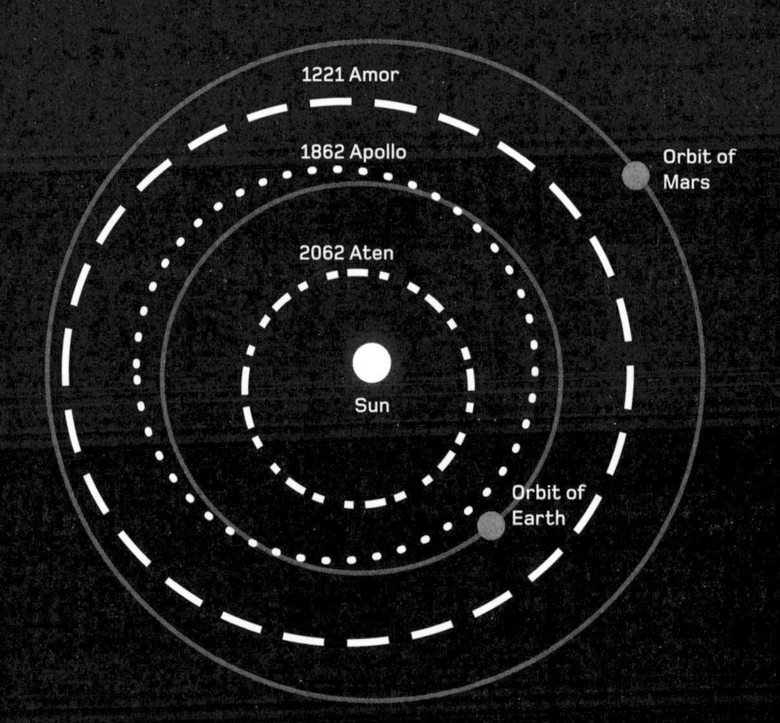

Orbits of prototype NEO asteroids

1 Ceres

Ceres was the first asteroid to be discovered (in 1801, by Italian astronomer Giuseppe Piazzi). Although now technically classified as a dwarf planet on account of its size (a near-perfect sphere some 945 km or 587 miles across), it still bears a 'minor planet number' – 1 – assigned according to a scheme invented in the mid-19th century. Ceres follows an elliptical orbit between 2.6 and 3.0 AU from the Sun, putting it beyond the 'frost line' of the solar system where ice is able to persist on the surface. In 2014, the infrared Herschel Space Observatory also discovered a thin atmosphere of water vapour that must be constantly replenished by sublimation from surface ice reservoirs. NASA's Dawn mission entered orbit around this small world in 2015, and sent back the first detailed images of its surface. Although heavily cratered, the landscape has surprisingly low relief – further evidence that the crust is rich in water ice with a tendency to 'relax' over time. Dawn also discovered distinctive bright spots inside some craters that appear to be salt deposits. One theory is that these mark locations where briny water from a subterranean ocean seeps to the surface.

4 Vesta

Despite being only the third-largest object in the asteroid belt (with a diameter of 560 km/348 miles), asteroid 4 Vesta is the only one bright enough to be seen from Earth with the naked eye. With an orbit averaging 2.4 AU from the Sun, Vesta is very different from the more distant icy Ceres. It might be large enough for gravity to pull it into a sphere (meriting the designation dwarf planet) were it not for a huge impact crater, called Rheasilvia, that has gouged a huge depression at the south pole and given the asteroid a mushroom-like shape.

NASA's Dawn mission spent a year orbiting Vesta in 2011/12, and confirmed earlier suspicions that it has a complex geological history. A covering of bright volcanic rock shows that Vesta was once hot enough to develop a layered, planet-like structure of crust, mantle and iron-rich core. Many similar asteroids may once have existed, before being ejected from the solar system or shattered in collisions. Their fragments are linked to distinct families of iron and 'stony' meteorites found on Earth.

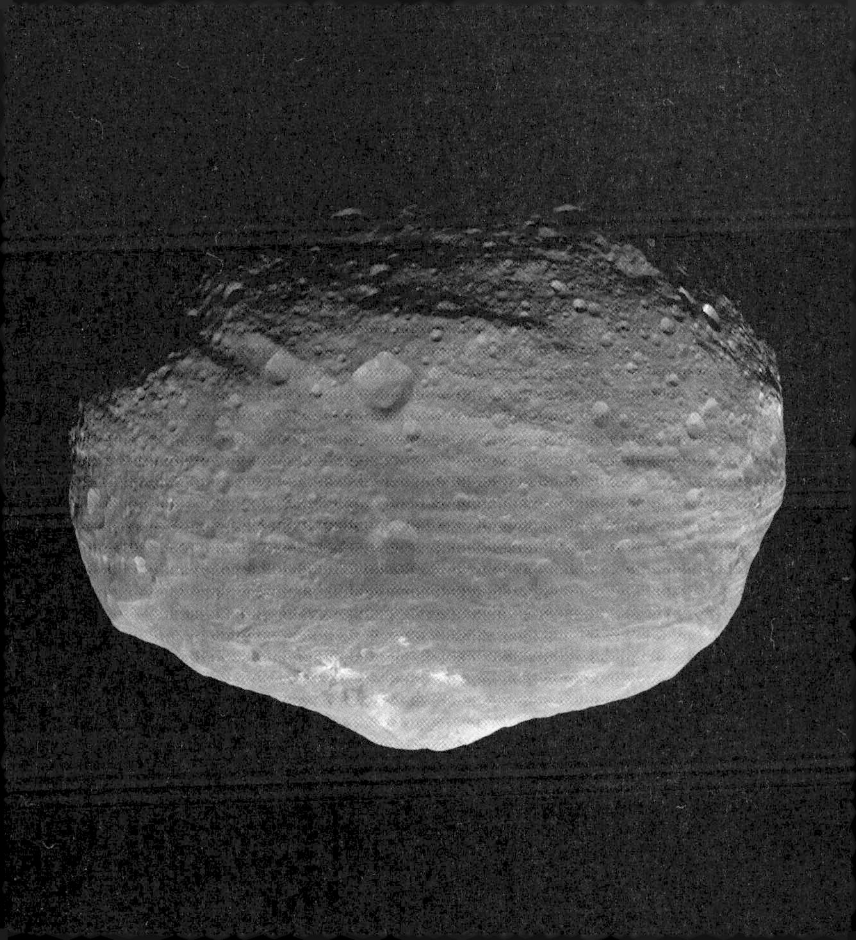

951 Gaspra

In 1991, the Jupiter-bound Galileo probe provided the first-ever close-up views of an asteroid during its brief flyby of 951 Gaspra. This misshapen lump of rock, some 20 km (12.4 miles) long, provided a number of surprises that make sense in the light of more recent thinking about the history of the asteroid belt. Where most astronomers had expected to see an ancient, heavily cratered surface, they found instead a surprisingly smooth one with several flat planes and only a light peppering of impacts. These suggest that Gaspra's current surface has only been exposed for between 20 and 300 million years. In fact, it's now thought that Gaspra is a member of a large asteroid grouping, known as the Flora family. Distinguished by similar orbits and silicate-rich mineralogy, all these asteroids are thought to have originated from the shattering of a relatively large object about 200 million years ago. The largest surviving fragment is the 140-km (87-mile) asteroid 8 Flora. About one third of all main-belt asteroids are now understood to form distinct families, and such occasional collisions are thought to have played a key role in the belt's evolution.

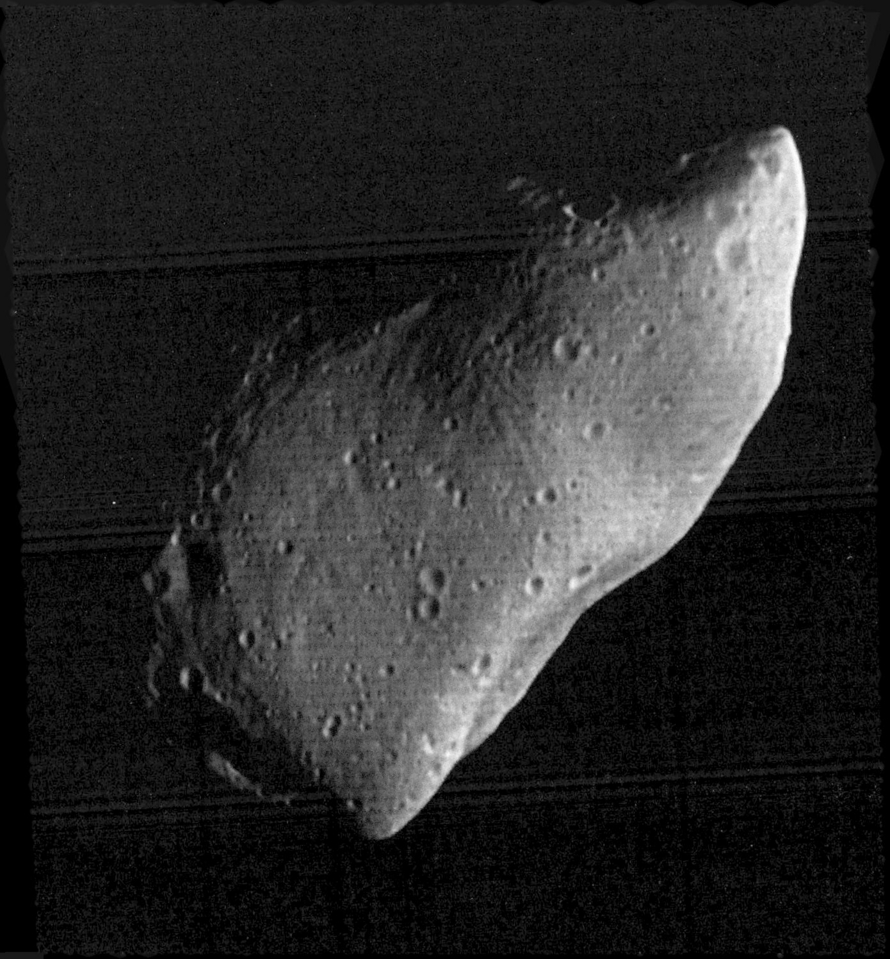

243 Ida and Dactyl

Following its brief encounter with 951 Gaspra (see page 172), NASA's Galileo probe moved on to fly past a significantly larger asteroid, 243 Ida. This irregularly shaped lump of rock, some 54 km (33.6 miles) long, spins on its axis every 4.6 hours, allowing Galileo's cameras to capture images covering most of its surface.

Ida's surface shows heavier cratering (and larger individual craters) than Gaspra, indicating that its surface is significantly older. It's now thought to be a member of the Koronis family, an asteroid group that originated in the breakup of a parent body about one billion years ago. Like Gaspra, Ida is classed as an S-type asteroid (a classification based on its colour and brightness as observed from Earth). Galileo's instruments confirmed for the first time that such asteroids are the primary source of stony 'ordinary chondrite' meteorites found on Earth. Flyby images also revealed that Ida is orbited by a small moon, 1.4-km (0.9-mile) Dactyl, which probably originated as a fragment chipped off during an impact on the larger asteroid.

253 Mathilde

Asteroid 253 Mathilde is a member of the inner asteroid belt, far from spherical but with an average diameter of 53 km (33 miles). It follows an orbit that takes it between 1.95 and 3.35 AU from the Sun and, in 1997, was targeted by the Near-Earth Asteroid Rendezvous (NEAR-Shoemaker) space probe on its way to Eros (see page 190). Mathilde proved to be a rough-hewn lump of rock, with several sharp-edged facets and a very dark surface (hence its primary features are named after Earth's major coal fields). It rotates only very slowly – once every 418 hours, while most other asteroids rotate in just a few hours unless their spin has been slowed by the influence of a satellite. Mathilde's dark rocks are similar to carbonaceous chondrite meteorites (see page 114), but it has much weaker gravity than might be expected for its size, suggesting that its interior must contain large voids. Mathilde's low gravity may explain its appearance – the major impacts it has endured would have thrown off debris with enough energy to escape the asteroid entirely, rather than settling in an ejecta blanket and smoothing the surface.

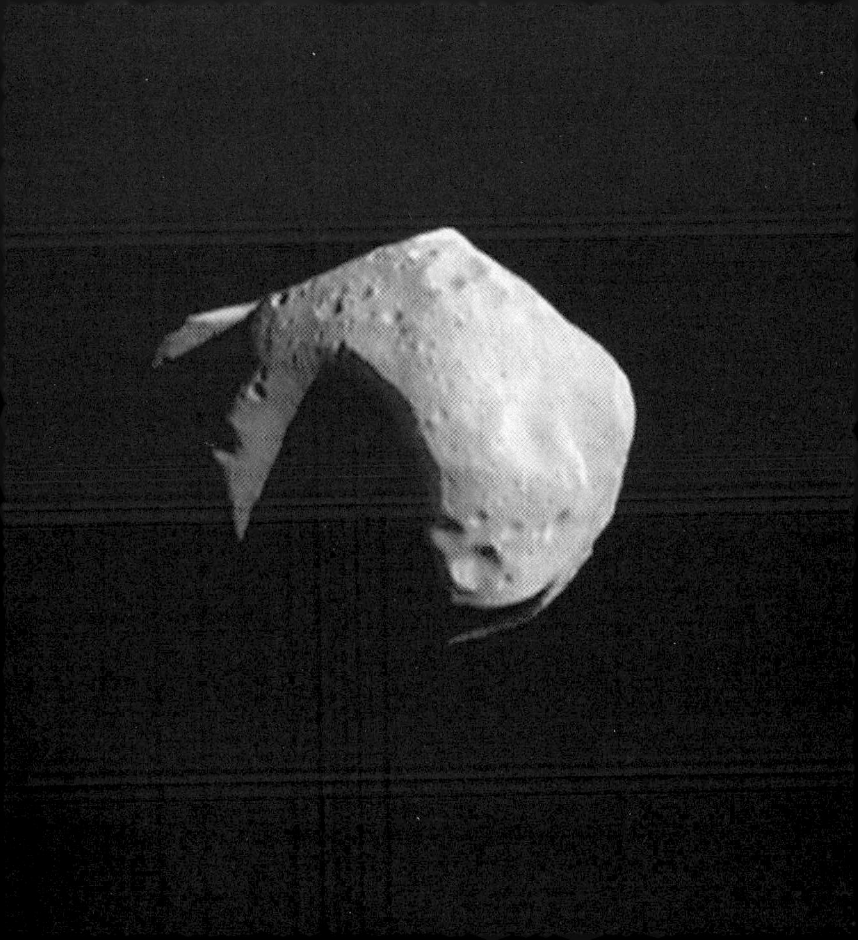

25143 Itokawa

Discovered in 2008 by the Lincoln Near-Earth Asteroid Research (LINEAR) project, 25143 Itokawa is a small asteroid in an Apollo-type orbit around the Sun (see page 166). Although not likely to threaten Earth in the foreseeable future, it is nevertheless classed as a potentially hazardous object. Itokawa's main claim to fame, however, is as the target of the Japanese Space Agency's Hayabusa space probe. Launched in 2003, Hayabusa reached Itokawa in 2005, collecting dust grains from the surface that were returned to Earth five years later. Hayabusa's cameras revealed an elongated asteroid with dimensions of 535 × 294 × 209 m (1755 × 965 × 686 ft) and a surface strewn with boulders and surprisingly lacking in impact craters. Itokawa's low density suggests a cavity-filled interior, and the dust retrieved by Hayabusa proved surprisingly young, having been exposed on the surface for about eight million years. All this suggests that Itokawa is a so-called 'rubble pile' asteroid, formed from fragments of earlier asteroids loosely bound together by their weak gravity.

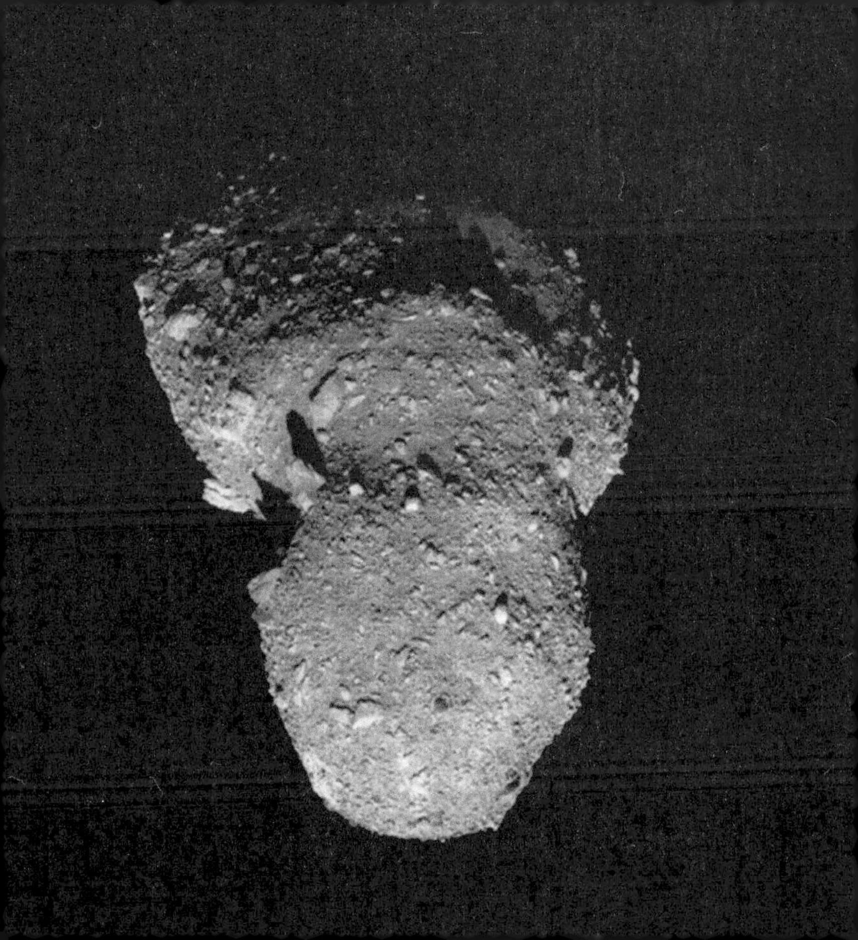

99942 Apophis

Despite a diameter of a mere 370 m (1,214 ft), 99942 Apophis is among the most intensively studied asteroids, thanks in large part to its potential to threaten Earth. A few months after its discovery in June 2004, the asteroid made a relatively close approach to Earth, which allowed its orbit to be properly calculated. This confirmed that Apophis will pass much closer to Earth (around 31,200 km/19,400 miles) on 13 April 2029, and raised the possibility of the asteroid hitting a small gravitational sweet spot or 'keyhole' that would divert its orbit and put it on course to strike Earth in 2036.

Fortunately, further observations in 2006 and 2013 confirmed that Apophis will miss the keyhole and instead have its orbit disrupted from that of an Aten-type NEO to an Apollo-type (see page 166). Nevertheless, the 2029 close approach will still be a salutary reminder of our planet's vulnerability to impact hazards, with the asteroid clearly visible to the naked eye from Europe, Africa and western Asia as it passes over Earth's night side.

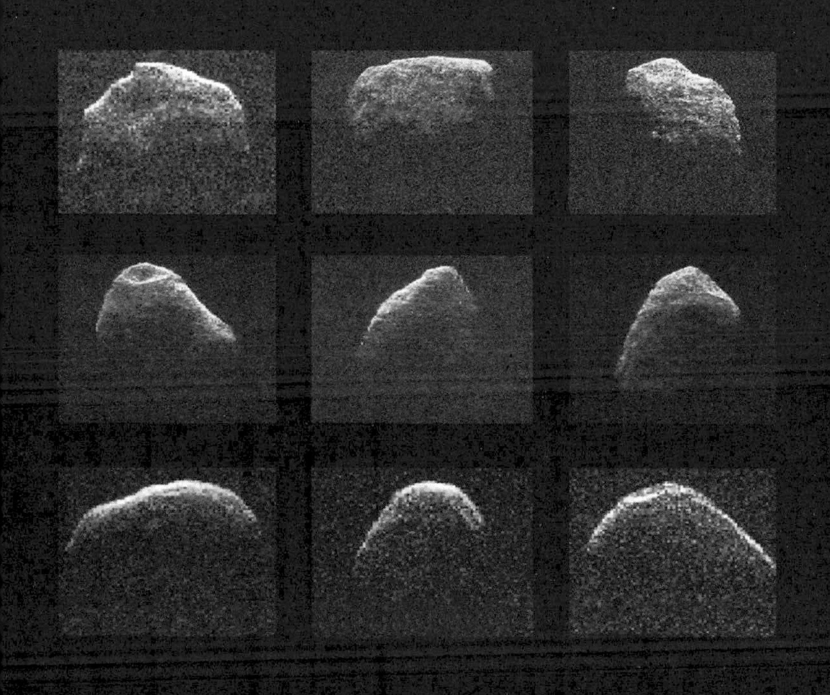

Radar images of Apophis, captured during its close approach to Earth in 2012

3200 Phaethon

Discovered in 1983, 3200 Phaethon is a roughly spherical asteroid with a diameter of around 5.8 km (3.6 miles), a dark surface and hints of an intriguing past. Phaethon's highly elliptical Apollo-type orbit (see page 166) carries it between 2.40 AU at aphelion and 0.14 AU at perihelion – well inside the orbit of Mercury. At the time of discovery, Phaethon's approach to the Sun was the closest of any known asteroid, and its orbit seemed more like that of a comet. The resemblance deepened when astronomers realized Phaethon's orbit matches that of the Geminid meteors – a stream of shooting stars that cross Earth's orbit every December.

Phaethon, therefore, is almost certainly an exhausted comet – albeit one with a significant rock component that has outlasted the loss of its ice. In 2009 and 2012, NASA's STEREO solar observatory satellites imaged Phaethon near the perihelion of its 524-day orbit and found signs of a 'dust tail' still being ejected from its dark surface.

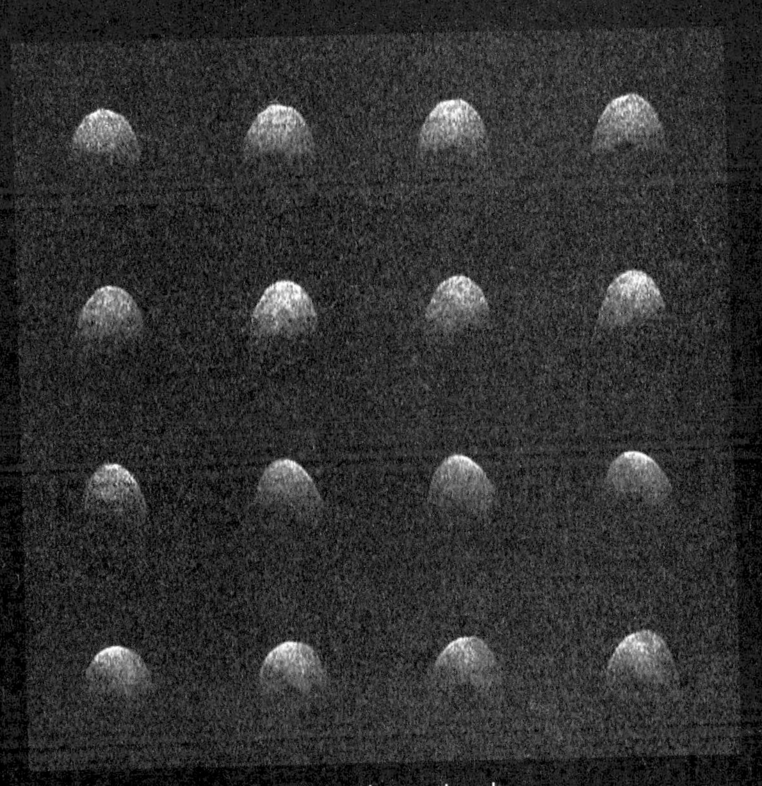

A sequence of radar images of Phaethon, captured during its 2007 close approach to Earth

592 Scheila

Although the main-belt asteroids are relatively well separated in space, their sheer numbers make collisions inevitable – and in 2010, astronomers apparently captured images of just such an event. 592 Scheila has an average diameter of 57 km (35 miles), and follows a markedly elliptical orbit between 2.45 and 3.41 AU from the Sun. It is thought to be a 'T-type' asteroid – one of a group of poorly understood dark-red worlds found in the inner asteroid belt and thought to lack water in their crusts.

When 2010 survey images revealed that Scheila had apparently developed a fuzzy halo or 'coma', therefore, it came as something of a surprise. Such comet-like behaviour is not unknown in asteroids from the icy outer belt, but would be unexpected from a dry asteroid like Scheila. Further studies confirmed this, showing that the coma was not composed of gas particles as a comet's would be. Instead, astronomers concluded that Scheila had been struck by a small object about 35 m (115 ft) across, flinging a substantial cloud of dust into a loose orbit around it.

3753 Cruithne

A small object designated as minor planet 3753, Cruithne is best known for its sometime reputation as Earth's 'other' moon. In fact, while this 5-km-diameter (3 mile) asteroid does not orbit directly around the Earth, it does have an intimate relationship to our planet. Cruithne's average distance from the Sun is more or less identical to Earth's, but its path is slightly more elliptical, so that it goes from just inside Earth's orbit to just outside it. For some of the time, its year is a little shorter than Earth's, so that it slowly spirals through space ahead of our planet, moving further ahead with each loop. Eventually, the asteroid starts to 'catch up' with Earth from behind, until it comes within 15 million km (9.3 million miles). At this distance, tidal forces from Earth rob Cruithne of a little momentum, so that it starts to orbit more slowly and retreat from Earth. Finally, Earth 'catches up' with the asteroid, and this time tidal forces boost Cruithne's momentum, causing it to pick up speed. This perpetual back-and-forth through space around Earth is known as a 'horseshoe orbit'.

1 Cruithne orbits closer to the Sun, faster than Earth.

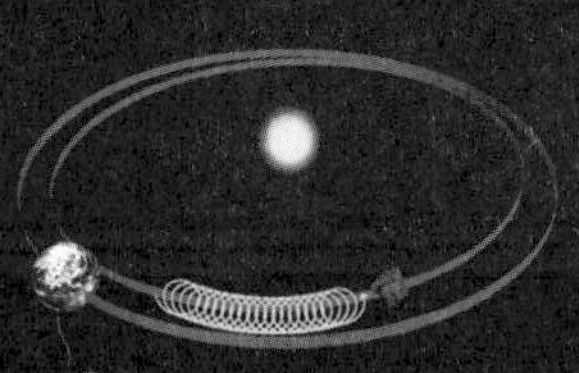

2 Cruithne slowly spirals ahead of Earth.

3 As Cruithne approaches Earth from behind, it is pulled into a wider orbit.

4 Now Cruithne orbits further out than Earth, and more slowly.

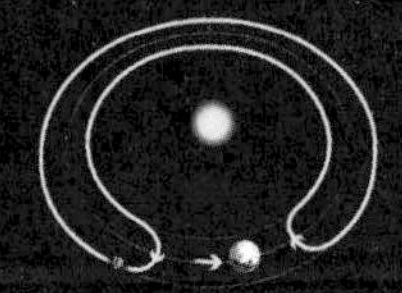

5 Overall 'horseshoe' path taken by Cruithne relative to Earth

2002 AA_{29}

In 2002, a systematic hunt for Near-Earth Asteroids discovered a 60-m-wide (200 ft), quarter-million-tonne asteroid lurking close to Earth's orbit around the Sun. Designated 2002 AA_{29}, this small rock follows a horseshoe orbit similar to that of Cruithne (see page 186). What's more, computer modelling suggests that every couple of thousand years it enters a brief temporary orbit around our own planet, to become a true 'second moon'.

Some astronomers have argued that this object's orbit is *so* similar to Earth's that it cannot be a mere captured stray from the asteroid belt. Instead, they suspect 2002 AA_{29} might have started its journey much closer to home, at one of the 'Lagrangian points' of the Earth–Moon system (gravitational sweet spots where an object can follow a stable orbit more or less indefinitely without either of the larger bodies disturbing it). If that's the case, it raises an intriguing possibility – could 2002 AA_{29} be a surviving chunk of debris thrown out from the 'Big Splash' impact thought to have created the Moon itself?

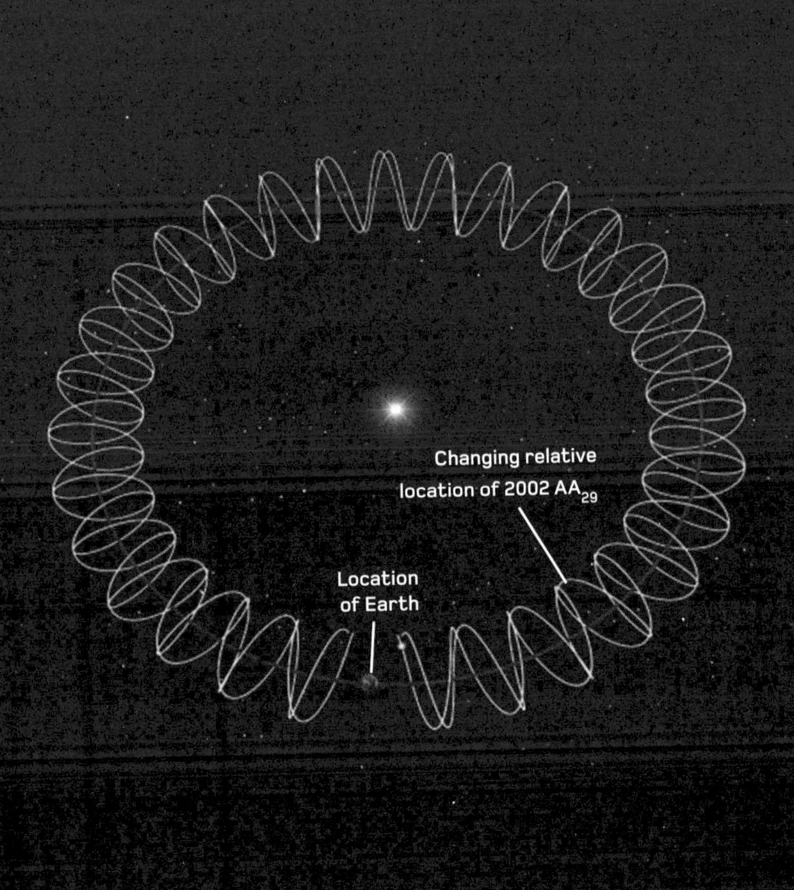

Changing relative location of 2002 AA_{29}
Location of Earth

433 Eros

Discovered by Gustav Witt in 1898, 433 Eros is one of many Near-Earth Asteroids that orbit inside the main belt. Circling the Sun once every 1.76 years, it is currently a Mars-crosser, though there is a strong chance that its orbit will eventually evolve to intercept Earth's.

As the ultimate destination of the Near-Earth Asteroid Rendezvous (NEAR-Shoemaker) space probe, Eros is among the most intensively studied asteroids in the solar system. The mission orbited the asteroid for 12 months, appropriately arriving on Valentine's Day 2000 and 'kissing' the surface with a final touchdown a year later. Eros's most distinctive feature is a saddle-shaped hollow in its convex side. The depression, known as Himeros, is almost certainly an ancient impact crater, and boulders up to 50 m (164 ft) across can be seen scattered around it. However, one of the biggest surprises about Eros was just how 'soft' its landscape appeared – presumably a result of erosion caused by endless bombardment from micrometeorites no bigger than dust grains.

Trojan asteroids

Since the discovery of 617 Patroclus and 624 Hektor (see page 194) in the early 20th century, astronomers have learned that Jupiter shares its orbit with at least 6,500 asteroids. These small objects avoid the disruptive influence of Jupiter's powerful gravity by orbiting near two of its 'Lagrangian points'. These are regions some 60 degrees behind and ahead of Jupiter, where the Sun's influence balances that of the giant planet and a stable orbit is possible. The asteroids in the two groups, named after rival Greeks and Trojans from the mythological Trojan war, probably reached their present orbits after being dispersed from their original locations by the migration of the giant planets proposed in the Nice model (see page 42). For some unknown reason, they contain a high number of binary asteroid pairs.

Today, any object that follows a similar orbit is also known as a Trojan – Earth, Mars, Uranus and Neptune have known Trojans of their own, while Saturn's satellites Tethys and Dione share their orbits with 'Trojan moons'.

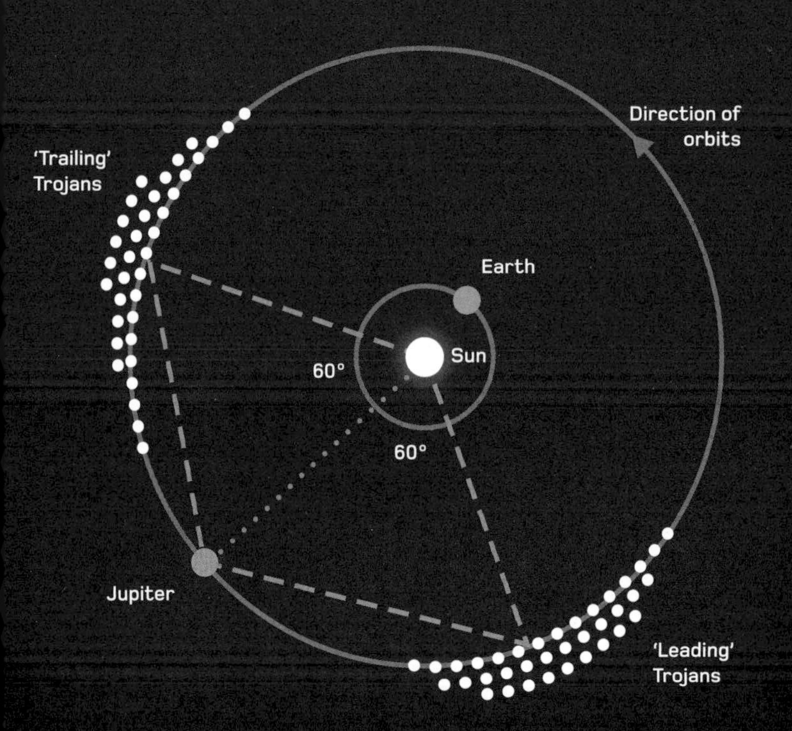
Orbits of the Trojan asteroids
Direction of orbits
'Trailing' Trojans
Earth
Sun
60°
60°
Jupiter
'Leading' Trojans

617 Patroclus and 624 Hektor

The two largest Trojan asteroids, Hektor and Patroclus, were discovered within months of each other in 1906 and 1907. Neither world has yet been visited by a space probe and they are hard to resolve through even powerful telescopes, but variations in their light have revealed a few details. In 2001, astronomers discovered that Patroclus is actually a binary system – a pair of oval asteroids orbiting each other at a distance of about 665 km (413 miles). The 'true' Patroclus has an average diameter of 113 km (70 miles), while the slightly smaller world, 104 km (65 miles) across, is now know as Menoetius. Their dark surfaces and relatively low masses suggest that they may be comet-like bodies from the outer solar system. Hektor, in contrast, is a single object with dimensions around 403 x 201 km (250 x 125 miles). Its two-lobed shape suggests that it may be a 'contact binary' – a pair of asteroids that collided and stuck together at some point in their past. Hektor is orbited by a 12-km-diameter (7.5 mile) moon, Skamandrios, and its dark-reddish surface matches a small number of other asteroids in the 'Hektor' family.

An artist's impression of Patroclus and its sibling world Menoetius

2015 BZ_{509}

In late 2017, astronomers announced the discovery of an elongated object called 'Oumuamua, retreating from a close encounter with the Sun. Its orbit proved it was not a true member of the solar system, but a fleeting visitor, an asteroid kicked out of orbit around another star millions of years ago. A few months later, scientists found evidence for another stray in the solar system – and this one is a permanent resident. The recently discovered asteroid 2015 BZ_{509} has a rare 'retrograde' orbit around the Sun, moving in the opposite direction to the planets in a Jupiter-crossing orbit. While the giant planet's gravity changes its orbit in the short term, the asteroid follows a long-term cycle that has been stable throughout the history of the solar system. Computer models suggest that 2015 BZ_{509} could not have achieved such an orbit starting from a 'normal' path – instead, the Sun must have swept it up (probably alongside several others) around 4.5 billion years ago, in a close encounter with another nascent solar system within its star-forming nebula.

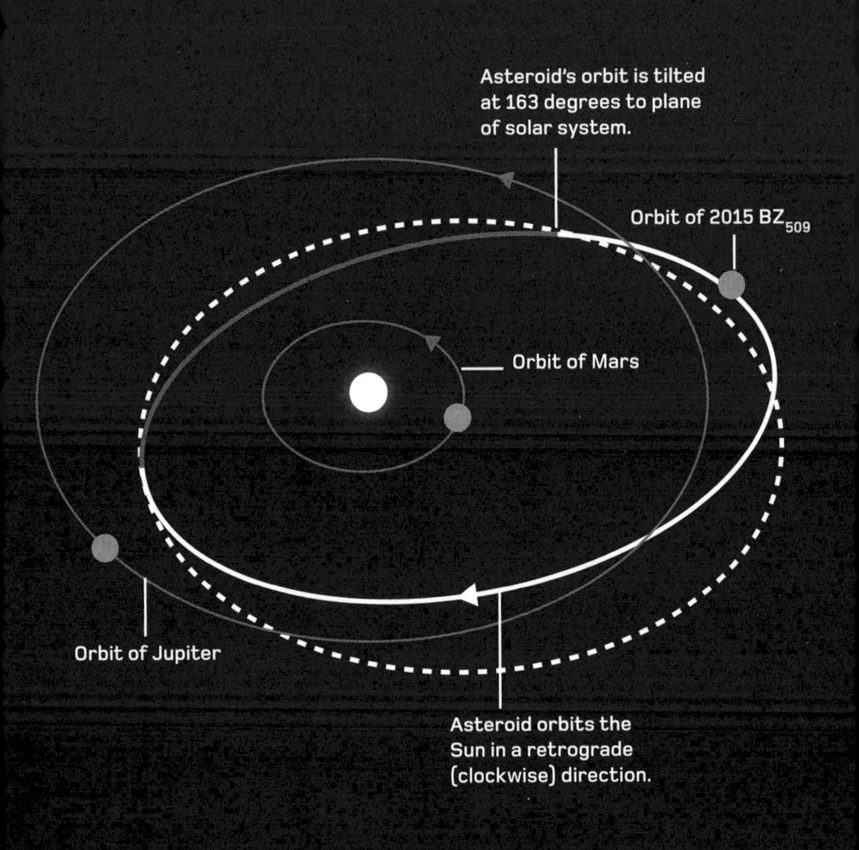

Asteroid's orbit is tilted at 163 degrees to plane of solar system.
Orbit of 2015 BZ_{509}
Orbit of Mars
Orbit of Jupiter
Asteroid orbits the Sun in a retrograde (clockwise) direction.

Jupiter

The solar system's largest planet by far, Jupiter is big enough to engulf all the other worlds with room to spare. Orbiting the Sun every 11.9 years at an average distance of 5.2 AU, it is a gas giant world – a vast ball of lightweight hydrogen in various forms, surrounding a hypothetical solid core with apparently 'blurred' edges. Everything about the planet is on an enormous scale – its average diameter of 139,822 km (86,881 miles) is 11 times that of Earth, and its mass 318 times greater. Unsurprisingly, therefore, this monster planet wields a huge influence over the rest of the solar system.

Jupiter's predominant features are its colourful weather systems, including multicoloured cloud bands that run parallel to the equator and turbulent, swirling storms of varying colours. As the fastest-rotating planet in the solar system (spinning on its axis in just 9 hours 55 minutes), even Jupiter's powerful gravity has difficulty holding onto fast-moving material at its equator, producing a pronounced equatorial bulge.

Inside Jupiter

Jupiter's colourful cloud bands and storms form the uppermost layer of a deep atmosphere that comprises the vast majority of the planet, with perhaps just a small solid core at the centre. Lightweight hydrogen is the dominant element. In the outer few thousand kilometres (Jupiter's true 'atmosphere') it takes the familiar form of gaseous molecules (H_2). However, deeper inside the planet it changes state as pressure increases, condensing at first into liquid hydrogen, and ultimately breaking down into a fluid metallic form whose swirling motions generate a powerful magnetic field (see page 206).

Although Jupiter's overall diameter has not changed much since its formation, its inner layers are in a state of gradual contraction, with denser materials sinking towards its centre. Similar processes are thought to take place inside all the giant planets, generating heat that helps to drive their weather systems – in Jupiter's case, this means that the planet emits more energy than it receives from the Sun.

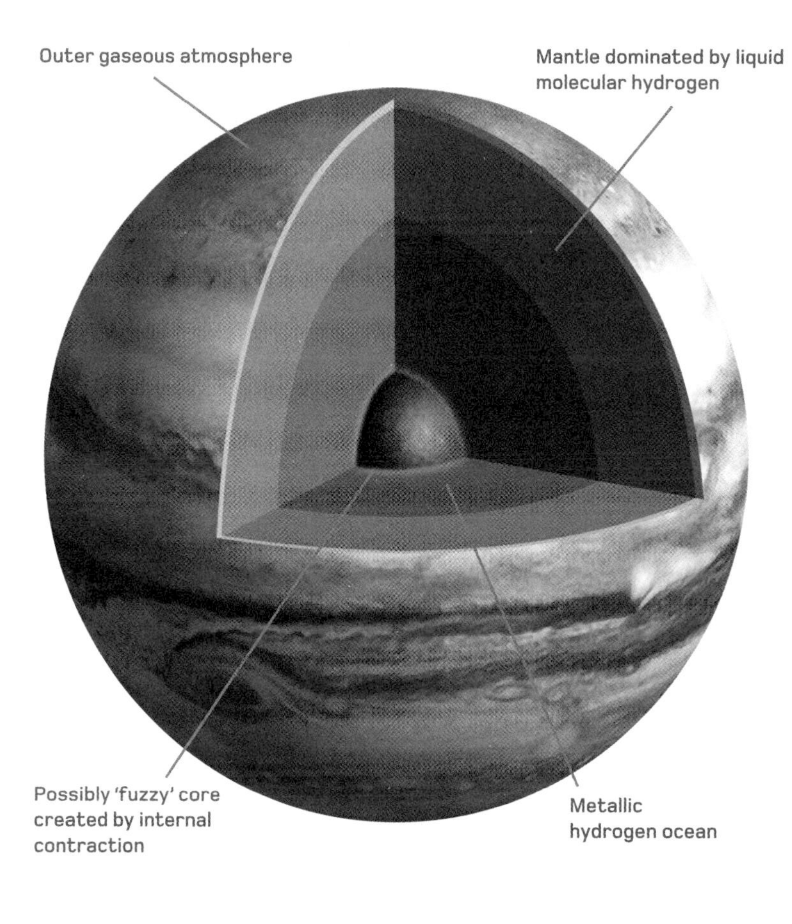
Outer gaseous atmosphere
Mantle dominated by liquid molecular hydrogen
Possibly 'fuzzy' core created by internal contraction
Metallic hydrogen ocean

Cloud belts and zones

Jupiter's contrasting cloud bands wrap around the planet, staying more or less parallel to the equator due to Jupiter's rapid 10-hour rotation. Bright, cream-coloured bands are called 'zones', while darker brown and blue stripes are known as 'belts'. Individual bands are identified according to their latitude.

Varying wind speeds in the atmosphere mean the belts and zones appear to move in opposite directions around the planet. The belt-and-zone terminology derives from the early assumption that the belts are overlaid on top of the zones, but space-probe flybys revealed that the real distribution of Jupiter's cloud patterns is exactly the opposite. Zones mark low-pressure regions within which clouds (in particular, those containing creamy-white ammonia crystals) condense at higher, colder altitudes. Belts, meanwhile, are regions where high-pressure forms 'clearings' that offer a view into the deeper layers of Jupiter's atmosphere, where the clouds are formed of more complex and colourful chemicals.

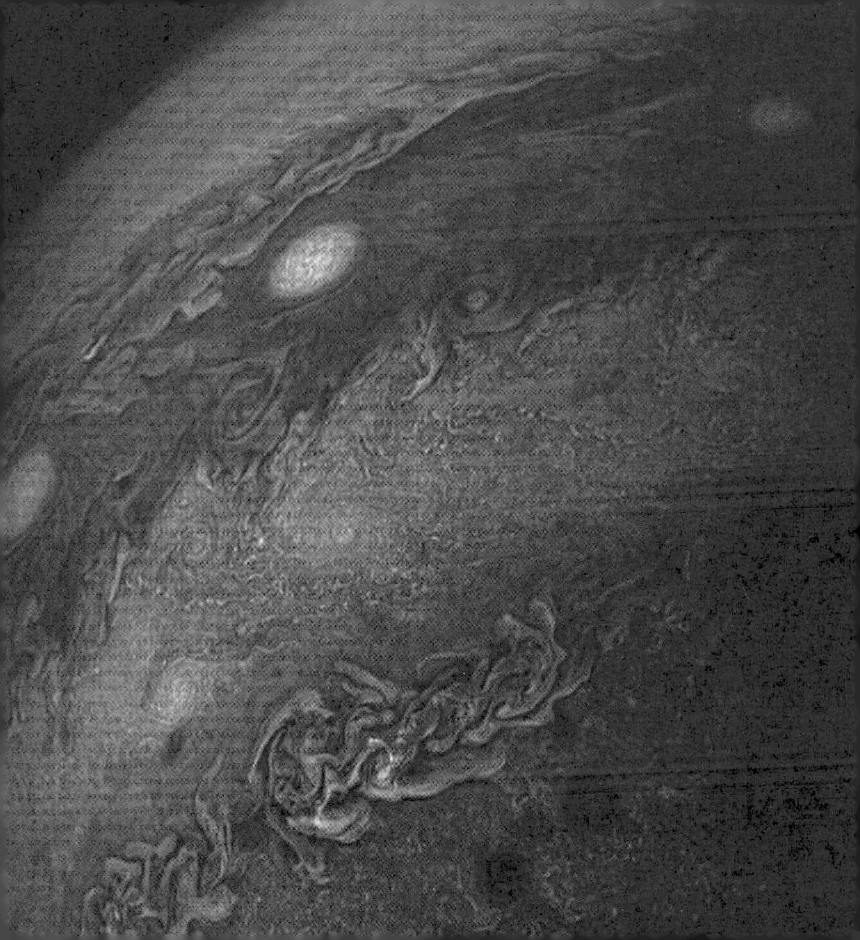

The Great Red Spot and other storms

Jupiter's most famous feature is a vast oval storm known as the Great Red Spot (GRS). Large enough to swallow the Earth, it swirls anticlockwise in tropical latitudes 22 degrees south of the equator, rotating once every six days. Space-probe images have confirmed that the GRS towers up to 8 km (5 miles) above the surrounding cloud levels, and it is usually interpreted as a vast high-pressure region that dredges up complex chemical material from deeper within the atmosphere. The spot has been observed continuously since 1830, and may well be much older than that, but it has been shrinking for much of the past century and may eventually disappear completely. Since the 1990s, astronomers have observed two other red spots in Jupiter's atmosphere. Oval BA or 'Red Spot Junior' formed from the merger of three smaller white-storm ovals in the late 1990s. It began to turn red in 2005 and has since continued to grow in size. In 2008, a so-called 'Baby Red Spot' appeared in Jupiter's southern cloud bands, but this was swiftly torn apart and cannibalized in a close encounter with the GRS.

Jupiter's magnetosphere

The vast, rapidly spinning sea of electrically charged, liquid metallic hydrogen that occupies much of Jupiter's interior creates a powerful magnetic field around the giant planet that is ten times stronger than that of Earth at its surface. However, this magnetosphere's strength and range are modified by an unusual interaction with the moon Io (see page 212). Active volcanoes on Io's surface eject large amounts of sulfur dioxide into a doughnut-shaped 'torus' that surrounds the moon's orbit. Solar radiation breaks the sulfur dioxide down into electrically charged sulfur and oxygen ions, creating a ring of electrically charged 'plasma' that effectively amplifies the magnetic field. The overall effect is to extend and flatten the magnetosphere, allowing Jupiter's magnetic influence to reach as far as the orbit of Saturn. Meanwhile, many of the trapped oxygen and sulfur particles rain down at high speeds into Jupiter's upper atmosphere above its poles, giving rise to intense aurorae (northern and southern lights) that emit radio waves, ultraviolet radiation and high-energy X-rays, as well as visible light.

Rings of Jupiter

Like all of the solar system's giant planets, Jupiter is surrounded by a system of rings. However, for such an impressive planet, the rings are perhaps one disappointing aspect – they amount to little more than a thin disc of dust particles in orbit around Jupiter. Divided into four distinct regions, they begin at around 20,000 km (12,500 miles) above the cloud tops and stretch as far as the orbit of the small satellite Thebe, some 154,500 km (96,000 miles) above the planet's equatorial cloudtops.

The rings are only visible when seen from Jupiter's night side and backlit by the Sun – as a result, they were only discovered when the retreating Voyager 1 photographed the system after its 1977 flyby. Two outer 'gossamer' rings and a relatively dense main ring each consist of dust ejected when micrometeorites strike Jupiter's small inner moons (Thebe and Amalthea for the gossamer rings, and both Metis and Adrastea for the main ring). The innermost 'halo' ring is thicker and consists of microscopic particles spread out by interactions with Jupiter's magnetic field.

Comet crashes

One of Jupiter's most important and under-appreciated roles is as the protector of the inner planets. Comets crossing its orbit on their way towards the Sun are prone to having their paths disrupted, and are often flung back into the outer reaches of the solar system (or even out of the Sun's grasp completely). Sometimes their orbits become circularized, so that encounters with Jupiter are more frequent and disruptive. In 1993, astronomers discovered that a close encounter with Jupiter had broken comet Shoemaker-Levy 9 into a string of smaller fragments doomed to collide with the gas giant the following year (see page 330). While seizing the opportunity to study the impact's effect on Jupiter's atmosphere, most regarded it as an infrequent chance event. However, the sighting of a 'scar' in Jupiter's clouds formed by another impact in 2009, and direct observations of further comet impacts since, show that objects hit Jupiter with surprising frequency, reducing the chances of them making damaging impacts on the far more vulnerable rocky planets.

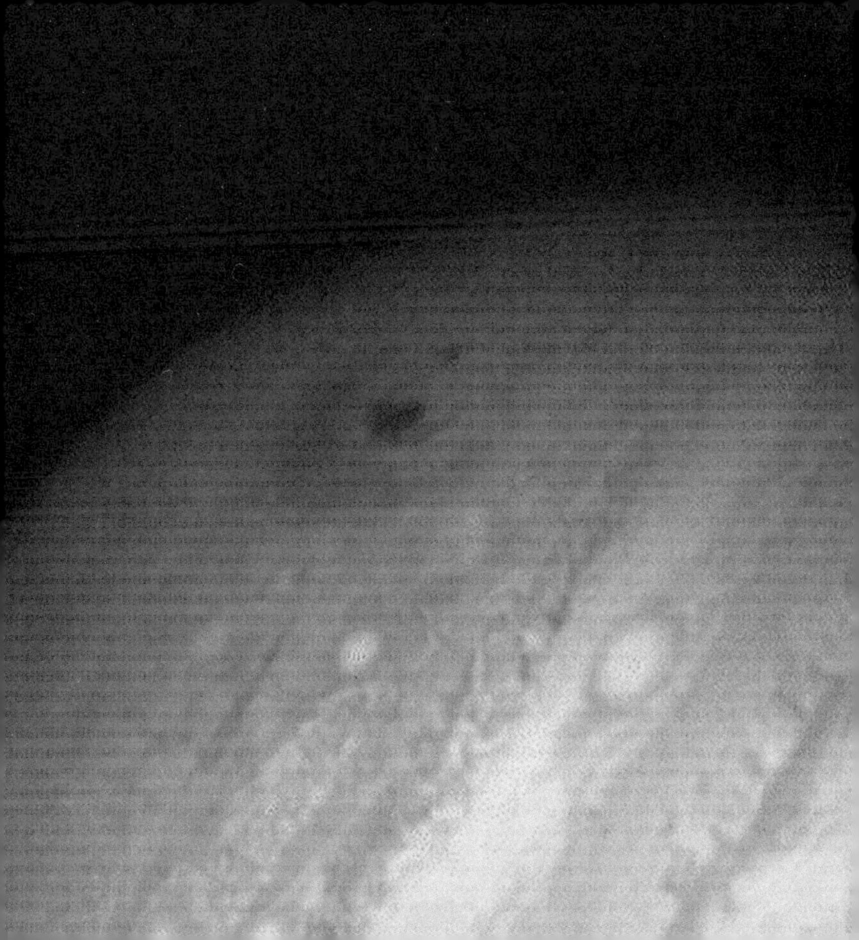

Io

Jupiter's vast family of satellites (69 at the latest count) is dominated by just four 'Galilean' moons – worlds that were first reported by Galileo Galilei when he turned his primitive telescope towards the giant planet in 1610. The innermost Galilean, Io, is a little larger than Earth's Moon and orbits Jupiter in just 42.5 hours – so close, that it is bombarded by radiation trapped in the planet's magnetosphere. Forced into a slightly elliptical orbit by the influence of its outer neighbours, Io suffers from tremendous tidal forces. Changes to the strength of Jupiter's gravity pull its interior in various directions and generate huge amounts of heat as Io's sulfur-rich rocks grind past each other. The rocks melt easily at relatively low temperatures, generating reservoirs of underground magma that make Io the most volcanically active world in the solar system. Its surface is constantly being changed and reshaped by new eruptions, but a survey by the Galileo space probe in the late 1990s identified more than 200 volcanic craters or calderas more than 20 km (12.5 miles) across.

Pele volcano

Io's volcanic activity takes various forms. Plumes of sulfur dioxide above the planet were first discovered during the Voyager 1 space-probe flyby of 1979. These enormous fountains of molten material resemble Earth's geysers; as they escape into Io's airless skies, they rapidly boil. The first of eight plumes to be found in Voyager 1 images was traced to a region that was subsequently named Pele, after the Hawaiian volcano goddess.

Pele itself is a volcanic crater or 'patera' with dimensions of 30 x 20 km (19 x 12 miles). The crater floor has several flat plains at varying levels formed in different phases of activity, and a lava lake at one end (the origin of the intermittent plume). Temperature measurements by space probes suggest that the lava reaches temperatures around 1,250°C (2,280°F), indicating that it is molten silicate rock similar to Earth's lava, rather than cooler sulfurous material. Crystals of sulfur, meanwhile, condensing out of the plume and settling back on the surface, form a reddish-brown ring around the patera.

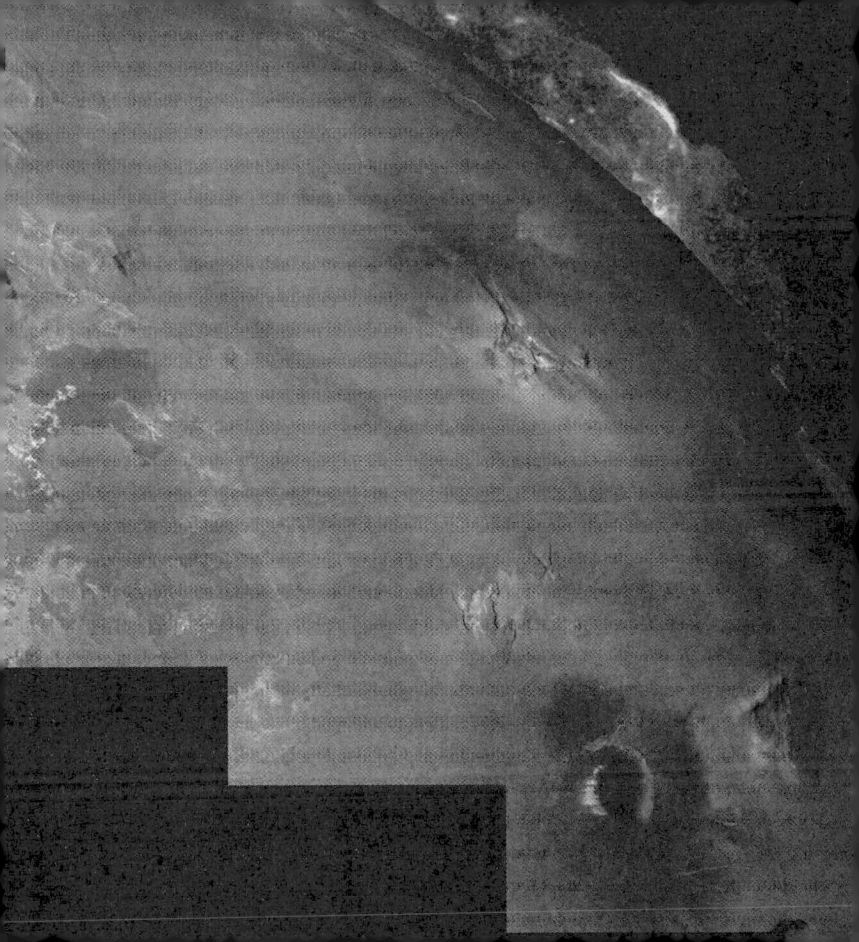

Europa

The second of Jupiter's major 'Galilean' moons, Europa is also the smallest, with a diameter of 3,122 km (1,940 miles) – slightly less than Earth's Moon. Its smooth, brilliant-white surface seems to have little in common with the colourful volcanic landscape of Io, but the moons are more alike than appearances suggest. Being in orbit between Jupiter and Io on one side and the much larger moon Ganymede on the other renders Europa vulnerable to the same tidal heating that shapes Io. In Europa's case, however, the presence of large volumes of water ice means that the heat has a different effect, warming a deep ocean of liquid water that is hidden from view (and protected from boiling into airless space) by a thick icy crust. Enhanced-colour images reveal that this crust is criss-crossed with pale, reddish-brown streaks, while some areas resemble the pack ice seen at Earth's polar caps. Most planetary scientists agree that these features are caused by the crust slowly rearranging itself over time, as mineral-laden water and warmer ice well up from beneath.

Pwyll crater

Only a handful of impact craters have been found on Europa, among which Pwyll is one of the youngest and best defined. Its central crater is about 26 km (16 miles) in diameter, while a surrounding spray of ejecta material extends across hundreds of kilometres, overlaying everything else within range. Pwyll's dark crater floor is probably formed from impure, mineral-laden ice that welled up from the ocean below. Its ejecta, in contrast, is pure water ice, vaporized out of the moon's icy crust as the crater formed, only to condense rapidly back into snow as it fell back to the surface.

Craters on Europa seem to disappear slowly with age, losing their definition on a relatively short timescale (perhaps thousands of years) as the icy material of the crust slumps and flattens out, until only a ghostly remnant survives. As a result of slow flowing of the surface, Europa's crust is so smooth that it is often compared to a pool ball; if this moon were the size of Earth, it would still have no hills higher than 200 m (660 ft).

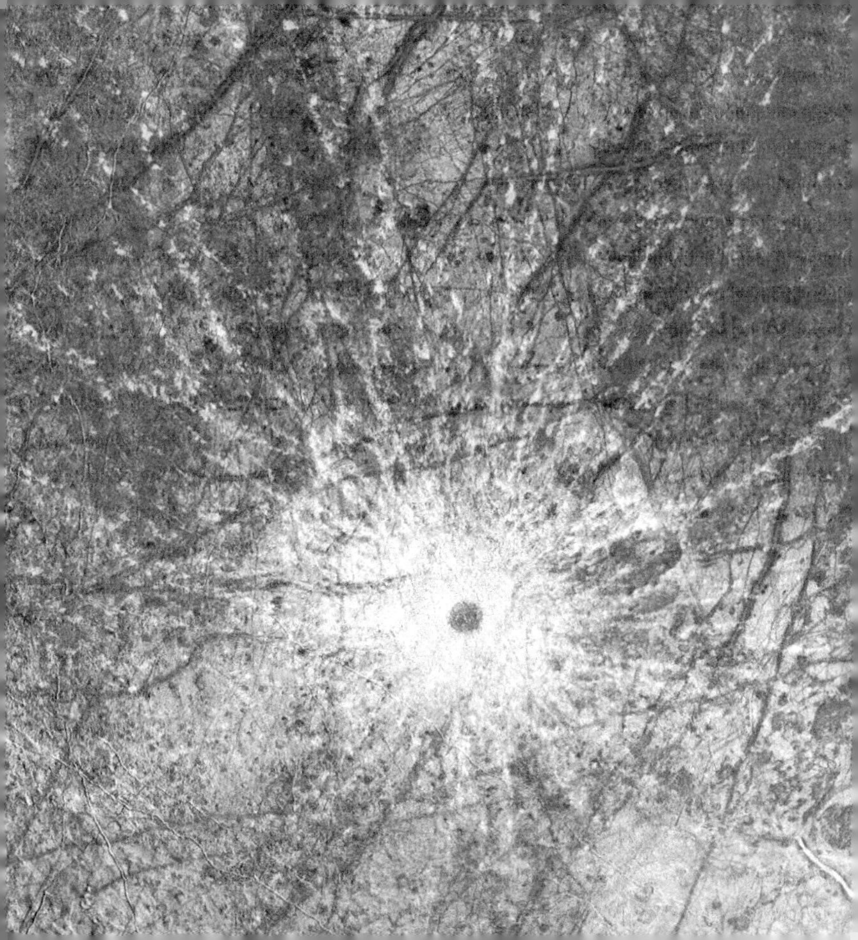

Europa's ocean

The presence of an ocean beneath Europa's icy crust was confirmed in the 1990s, when the Galileo space probe discovered the magnetic field it produces through interaction with Jupiter's own magnetosphere. Europa's water layer is thought to be about 100 km (60 miles) deep, with the upper 10–30 km (6–19 miles) composed of various forms of ice. The upper crust is frozen rock hard by surface temperatures around –160°C (–256°F), but underlying 'warm' ice that pushes its way towards the surface brings clues to the nature of the ocean beneath. For example, long surface streaks called 'lineae' are stained by colourful chemicals, such as magnesium sulfate, that probably originated from volcanic vents on the sea floor (similar to those that belch minerals into Earth's oceans). This raises the intriguing possibility that, like Earth's own 'black smokers', Europa's sea-floor vents could give rise to their own uniquely adapted forms of life. The discovery of water plumes above Europa, similar to those ejected by Saturn's moon Enceladus, may offer future space probes a way to investigate the ocean's chemistry without a technically challenging drilling mission.

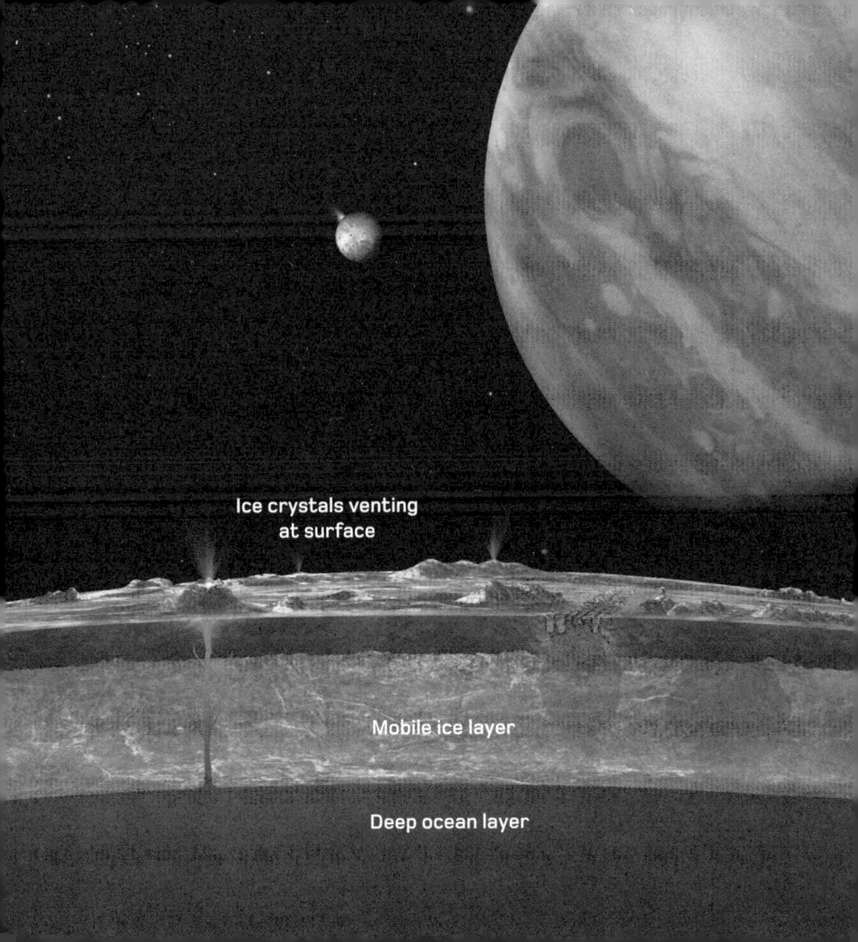

Ice crystals venting
at surface
Mobile ice layer
Deep ocean layer

Ganymede

The third and largest of Jupiter's Galilean moons, Ganymede is the biggest moon in the solar system. With a diameter of 5,269 km (3,274 miles), it's significantly larger than the planet Mercury. Despite its size, Ganymede has no significant atmosphere, but a substantial magnetic field suggests the presence of a core that still has molten iron. Other magnetic effects, meanwhile, seem to reveal a saltwater ocean layer around 200 km (124 miles) below the surface.

The moon's frozen surface, made from a mix of ice and rock, has a mottled appearance in which dark, old and heavily cratered regions are separated by lighter, less cratered (and therefore more recently formed) areas. It seems that heat escaping from Ganymede's interior once drove an icy equivalent of Earth's plate tectonics, pulling older areas of the crust apart and allowing a fresh mix of rock and ice to well up and fill the gaps. This suggests that, while Ganymede is not subject to tidal heating today, things were different in the distant past.

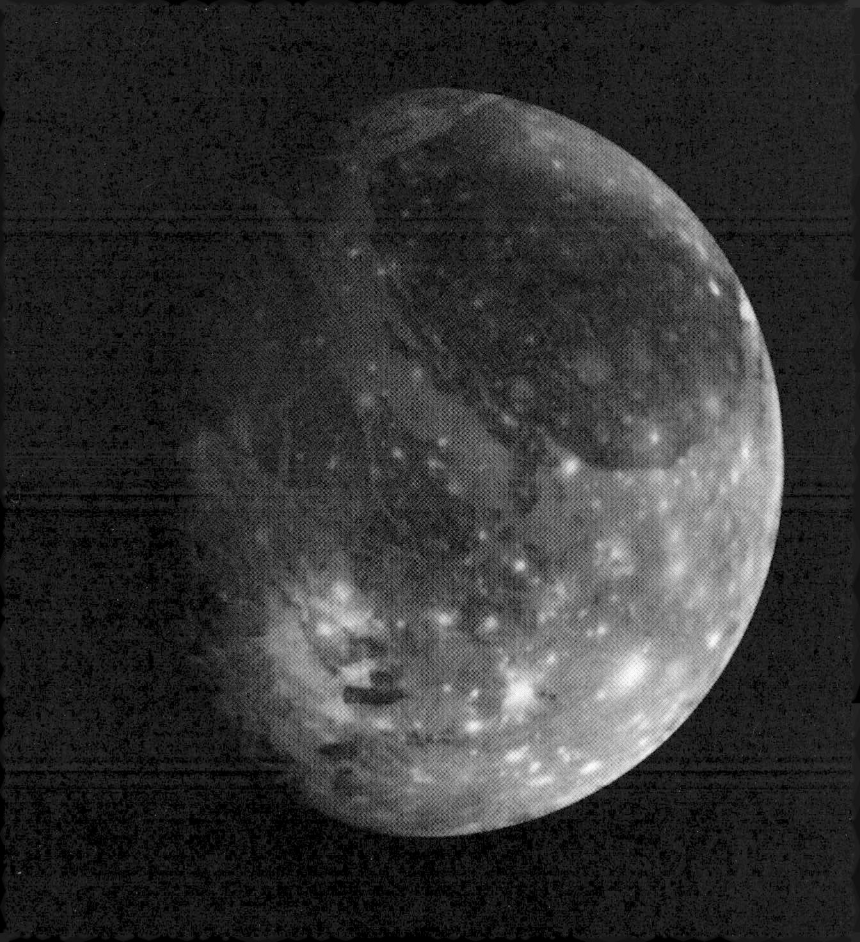

Tiamat Sulcus

In many places, the brighter segments of Ganymede's crust take the form of long, broad strips of parallel grooves and ridges, known as 'sulci'. The longest, such as Tiamat Sulcus, stretch for hundreds of kilometres across the giant satellite's landscape, forming ranges of rolling hills that may be up to 500 m (1600 ft) high, separated by parallel valleys a few kilometres wide.

Sulci are rather similar to the parallel ridges associated with the generation of new crust on Earth, and it's likely they have a similar cause. They seem to mark areas where new crust was created in the distant past. Convection in the moon's interior caused the original crust to crack into icy plates that slowly drifted in different directions, allowing fresh ice to well up through the cracks. The process evidently continued for a considerable part of Ganymede's history, with older sulci being subjected to the same process; Tiamat, for example, is divided in two by the narrower Kishar Sulcus.

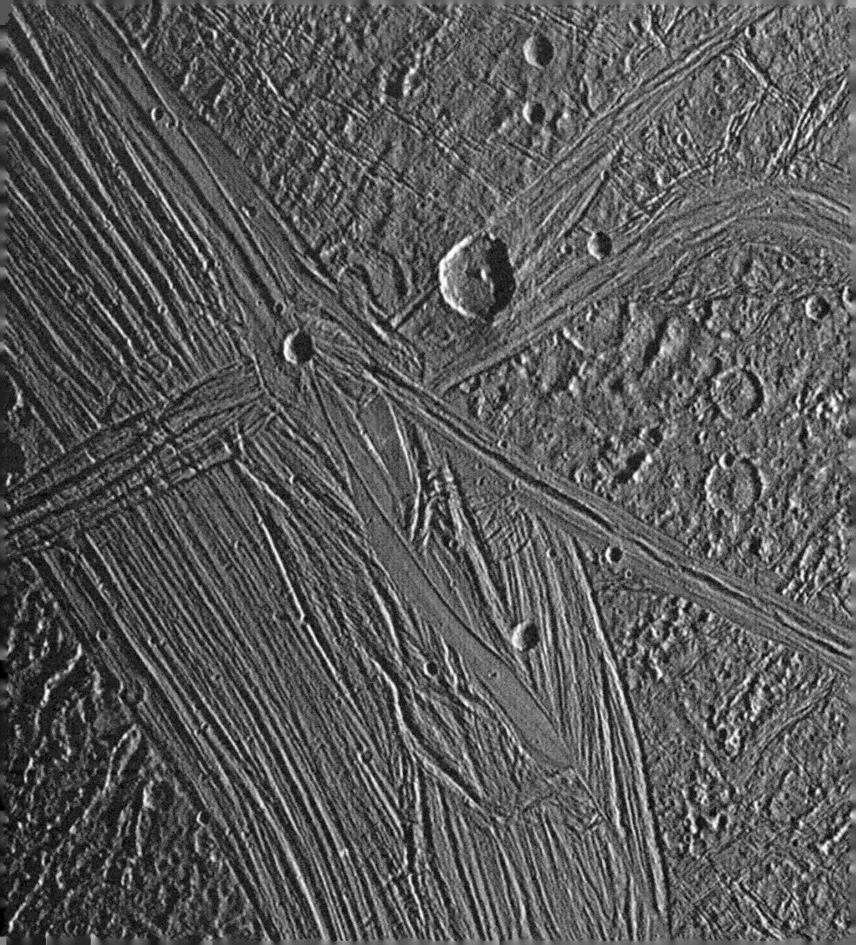

Callisto

The outermost of Jupiter's Galiean moons and the second largest (with a diameter of 4,821 km or 2,996 miles), Callisto has never suffered the tidal heating that helped to shape its inner neighbours. As a result, its surface has been stable since its formation, and it is now saturated with the impacts from more than four billion years' worth of objects pulled inward by Jupiter's gravity. Researchers believe it is probably the most cratered world in the solar system. Although the outermost crust is generally dark (a result of 'space weathering': bombardment by particles from the solar wind), impacts force out fresh ice from just below the surface, spraying bright rays of ejecta across the landscape.

Callisto's lack of geological activity suggests that its cold and icy interior never separated to form a distinct core. However, the moon's effect on Jupiter's magnetic field suggests there is at least one internal layer – a global salty ocean around 150 km (93 miles) deep, hidden beneath about the same depth of solid, icy crust.

Asgard and Valhalla

Callisto's dominant features are two enormous impact basins, known as Valhalla and Asgard. Valhalla is the larger of the two, with an overall diameter of about 1,900 km (1,180 miles). Asgard (opposite) is only slightly smaller at 1,600 km (1,000 miles) wide. Each basin consists of a relatively bright, flat central plain known as a 'palimpsest', surrounded by rings of concentric hills rather than a single defined crater wall. In places, the Sun's feeble, but still potent, heat causes ice to sublimate out of the rock-ice mix, weathering these hills into sharply defined, conical peaks. The palimpsests (a term used by historians to describe manuscripts that have been wiped clean and reused) are believed to have formed where each incoming meteorite smashed through the moon's dark outer crust and allowed brighter, relatively slushy ice to well up from beneath, eventually healing the central scars. Patterns of radial fractures that cross the surrounding mountain ranges, meanwhile, may have been caused by the crust flexing on top of the moon's deep underlying ocean as the basin formed.

Amalthea

In addition to its four giant Galilean moons, Jupiter has a host of other satellites. Of these, 61 outer moons are most likely objects captured into orbit around the planet, while four inner moons, orbiting close to Jupiter's rings, appear to be natural satellites, formed alongside the planet. Amalthea is the largest of these, and third in distance from Jupiter. Some 262 km (163 miles) long and 150 km (93 miles) wide, it is peppered with craters, the largest of which is the 90-km-diameter (56 mile) Pan. Amalthea also has one of the reddest surfaces in the solar system (one theory is that it sweeps up reddish sulfurous material that escapes from Io). Data from the Galileo space probe showed that Amalthea has a surprisingly low mass (suggesting that it is little more than an orbiting 'rubble pile'), while infrared studies have revealed the presence of hydrated minerals that are hard to explain if it formed at its current distance from Jupiter. It's thought, therefore, that Amalthea is all that survives of a large progenitor body that once orbited further out; when this was destroyed in an impact, some of the pieces reassembled themselves into this moon.

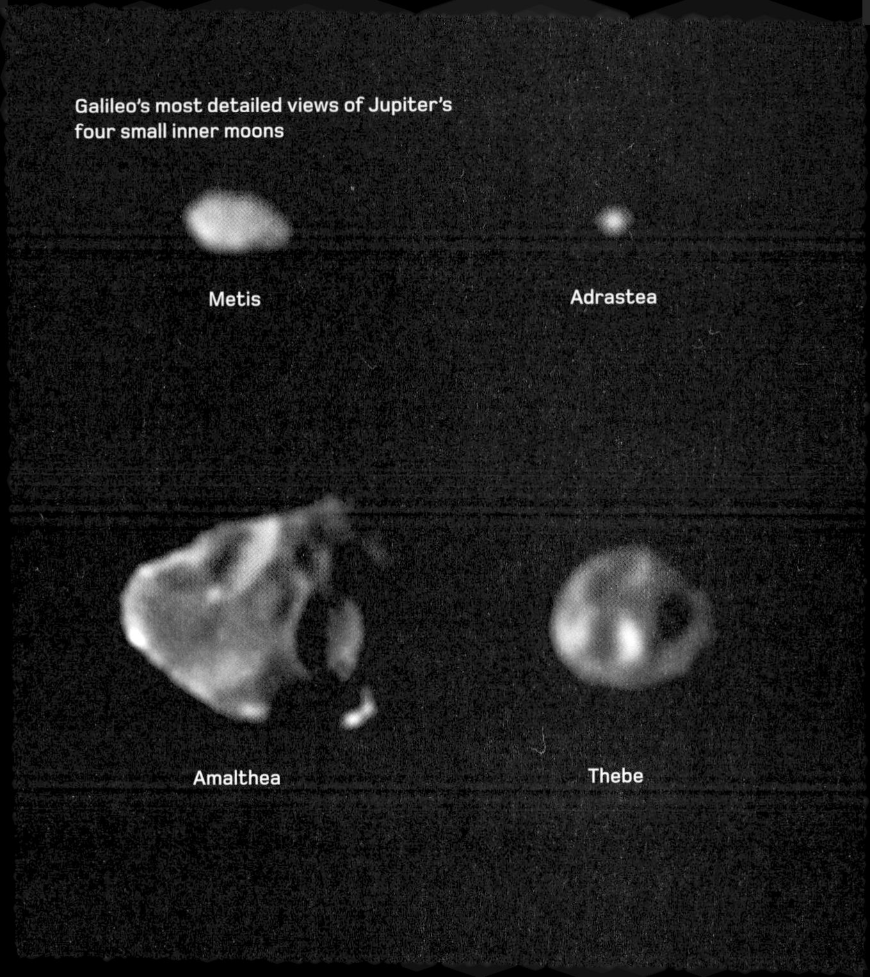

Galileo's most detailed views of Jupiter's four small inner moons

Metis

Adrastea

Amalthea

Thebe

Moons of Jupiter

Name	Diameter	Orbital period (days)*	Eccentricity (circular = 0)
Metis	60x40x34 km (37x25x21 miles)	0.29	0.00
Adrastea	20x16x14 km (12x10x9 miles)	0.30	0.00
Amalthea	250x146x128 km (155x91x80 miles)	0.50	0.00
Thebe	116x98x84 km (72x61x52 miles)	0.67	0.02
Io	3643 km (2263.7 miles)	1.77	0.00
Europa	3121.6 km (1939.7 miles)	3.55	0.01
Ganymede	5262.4 km (3269.9 miles)	7.15	0.00
Callisto	4820.6 km (2995.4 miles)	16.69	0.01
Themilessto	8 km (5 miles)	130	0.24
Leda	10 km (6.2 miles)	241	0.16
Himalia	170 km (105.6 miles)	251	0.16
Lysithea	24 km (14.9 miles)	259	0.11
Elara	80 km (49.7 miles)	260	0.22
Dia	4 km (2.5 miles)	287	0.25
Carpo	6 km (3.7 miles)	456	0.43
S/2003 J3	4 km (2.5 miles)	504 (R)	0.24
S/2003 J12	2 km (1.2 miles)	533 (R)	0.38
Euporie	2 km (1.2 miles)	553 (R)	0.16
S/2011 J1	4 km (2.5 miles)	581 (R)	0.30

* R = Retrograde orbit

Name	Diameter	Period*	Ecc.
S/2010 J2	4 km (2.5 mi)	588 (R)	0.31
S/2003 J16	4 km (2.5 mi)	595 (R)	0.27
S/2016 J1	3 km (1.9 mi)	604 (R)	0.14
S/2003 J18	4 km (2.5 mi)	606 (R)	0.12
Mneme	4 km (2.5 mi)	620 (R)	0.23
Euanthe	3 km (1.9 mi)	621 (R)	0.23
Orthosie	2 km (1.2 mi)	623 (R)	0.28
Harpalyke	4.4 km (2.7 mi)	623 (R)	0.23
Praxidike	6.8 km (4.2 mi)	625 (R)	0.22
Thyone	4 km (2.5 mi)	627 (R)	0.23
Thelxinoe	4 km (2.5 mi)	628 (R)	0.22
Ananke	20 km (12.4 mi)	630 (R)	0.24
Iocaste	5.2 km (3.2 mi)	632 (R)	0.22
Hermippe	4 km (2.5 mi)	634 (R)	0.21
Helike	8 km (5 mi)	635 (R)	0.16
S/2003 J15	4 km (2.5 mi)	668 (R)	0.11
S/2003 J9	2 km (1.2 mi)	683 (R)	0.27
S/2003 J19	4 km (2.5 mi)	701 (R)	0.33
Autonoe	4 km (2.5 mi)	715 (R)	0.20
Pasithee	2 km (1.2 mi)	716 (R)	0.29
Herse	4 km (2.5 mi)	717 (R)	0.28
S/2003 J4	4 km (2.5 mi)	723 (R)	0.20
S/2010 J1	4 km (2.5 mi)	723 (R)	0.32
Chaldene	3.8 km (2.4 mi)	724 (R)	0.24
Eurydome	3 km (1.9 mi)	724 (R)	0.26

Name	Diameter	Period*	Ecc.
Isonoe	3.8 km (2.4 mi)	726 (R)	0.26
S/2011 J2	4 km (2.5 mi)	727 (R)	0.39
Erinome	3.2 km (2 mi)	728 (R)	0.27
Kale	2 km (1.2 mi)	730 (R)	0.26
Aitne	3 km (1.9 mi)	730 (R)	0.26
Taygete	5 km (3.1 mi)	732 (R)	0.25
S/2017 J1	2 km (1.2 mi)	734 (R)	0.40
Carme	30 km (18.6 mi)	734 (R)	0.25
Cyllene	4 km (2.5 mi)	738 (R)	0.32
Hegemone	6 km (3.7 mi)	740 (R)	0.33
Kalyke	5.2 km (3.2 mi)	743 (R)	0.24
Pasiphae	36 km (22.4 mi)	744 (R)	0.41
Eukelade	8 km (5 mi)	746 (R)	0.27
Sponde	2 km (1.2 mi)	748 (R)	0.31
Megaclite	5.4 km (3.4 mi)	753 (R)	0.42
Callirrhoe	8 km (5 mi)	759 (R)	0.28
Sinope	28 km (17.4 mi)	759 (R)	0.25
S/2003 J5	8 km (5 mi)	760 (R)	0.21
S/2003 J23	4 km (2.5 mi)	760 (R)	0.31
Aoede	8 km (5 mi)	762 (R)	0.43
Arche	3 km (1.9 mi)	763 (R)	0.33
Kallichore	4 km (2.5 mi)	765 (R)	0.26
S/2003 J10	4 km (2.5 mi)	767 (R)	0.21
Kore	4 km (2.5 mi)	779 (R)	0.33
S/2003 J2	4 km (2.5 mi)	983 (R)	0.38

Saturn

Like Jupiter, Saturn is a gas giant with a huge atmosphere wrapped around a small solid core. The sixth planet from the Sun and the most distant visible to the naked eye, it orbits at an average distance of 9.6 AU. A tilted axis similar to Earth's gives rise to a comparable cycle of seasons – albeit stretched over the 29.5 Earth years it takes to complete a single orbit.

Aside from its spectacular rings (see pages 242–51), Saturn appears at first to be placid in comparison to Jupiter, with an orderly pattern of cream-toned cloud bands. However, the two worlds are far more alike than such appearances suggest – much of Saturn's activity is simply 'muted' to distant observers by creamy ammonia clouds that condense at high altitudes in its atmosphere. The planet's average diameter is 116,464 km (72,367 miles) – some 83 per cent of Jupiter's, but it has less than one-third of that planet's mass and, consequently, low density and gravity. Coupled with the planet's rapid 10.5-hour rotation, this creates a noticeable bulge around Saturn's equator.

Inside Saturn

As a gas giant, Saturn has a fairly similar internal structure to Jupiter. It consists primarily of hydrogen and helium, with a gaseous atmosphere transforming into a liquid mantle at high pressures about 1,000 km (620 miles) beneath the visible surface. Further in, liquid molecular hydrogen breaks down into a sea of 'liquid metallic' hydrogen that stretches down to a central core perhaps twice the size of Earth. Overall, however, the planet's average density is less than that of water – Saturn would float if immersed in a big enough ocean.

Currents running through the electrically conductive sea generate a magnetic field somewhat weaker than Jupiter's. Just as the Jovian field is boosted by interaction with material escaping from Io (see page 206), so Saturn's field is shaped and intensified by the influence of water vapour ejected from the small moon Enceladus into the doughnut-shaped E ring (see page 250). The magnetic field also channels trapped particles from the solar wind onto the poles, to create spectacular aurorae.

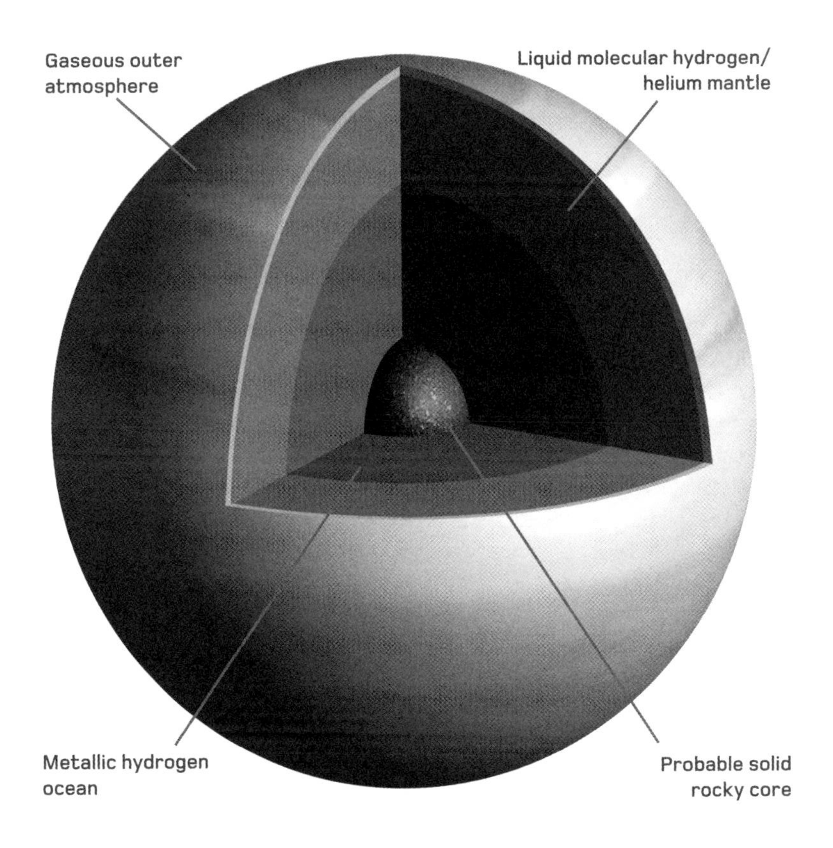
Gaseous outer atmosphere
Liquid molecular hydrogen/ helium mantle
Metallic hydrogen ocean
Probable solid rocky core

Saturnian storms

Saturn's most obvious weather features are bands of darker and lighter cloud that wrap their way around the planet, moving in opposite directions under the influence of prevailing winds. Lighter bands are defined as zones and somewhat darker ones as belts, but both are generally wider and less well-defined than the similar features on Jupiter (see page 202).

Although there is no semi-permanent storm to match Jupiter's Great Red Spot, certain regions of Saturn's atmosphere are prone to generating storms on a more or less predictable basis. At mid-southern latitudes, a region known as Storm Alley produces electrical storms deep within the atmosphere. Radio bursts from one particularly large example, called the Dragon Storm, were detected by the Cassini probe as it neared Saturn in 2004. A visually more impressive storm called the Great White Spot (opposite) recurs in the northern hemisphere roughly every 30 years, usually bursting into visibility around the height of the northern summer when the atmosphere is warmest.

The polar hexagon

Saturn's north pole is marked by a striking hexagonal structure of dark clouds, about 13,800 km (8,600 miles) along each side, with a whirlpool-like vortex embedded at its centre. One possible explanation for this strangely geometric feature is that its shape is created by the sharp boundary between faster- and slower-moving regions of Saturn's atmosphere. Careful measurement from NASA's Cassini probe suggests that, unusually, the hexagon is rotating at exactly the same rate as the planet's deep interior – viewed from that frame of reference, it has no significant directional winds (in striking contrast to the rest of the atmosphere). Wind speeds around the central vortex, however, are some of the strongest in the solar system, reaching around 1,800 km/h (1,120 mph). A prominent 'eye' at its centre, some 2,000 km (1,240 miles) across, is surrounded by a wall of clouds up to 75 km (47 miles) high. These strange features may be connected to Saturn's unusually warm poles – temperatures here can be up to 60°C (140°F) warmer than in regions close to the equator.

Saturn's rings

The ring system around Saturn is the largest and most complex in the solar system. Its prominent inner section extends across more than twice Saturn's own diameter, but is just 1 km (0.6 miles) thick at most. Seen from orbiting spacecraft, these razor-thin platters dissolve into countless narrow ringlets of different brightness and transparency, separated by distinct gaps.

Each ringlet is itself a stream of closely packed individual particles following near-perfect circular orbits above Saturn's equator. Small-scale disturbances in the rings are common, but the dynamics of the system naturally restore an orderly rotation. Any particular ring particle that is pushed into an elliptical or inclined orbit is much more likely to collide with its fellows, these collisions tending to cancel out its rogue movement. Astronomers have understood the basic nature of the rings since 1859, when Scottish mathematician James Clerk Maxwell showed that if the rings were solid structures, they would inevitably be torn apart by tidal forces.

A and B rings

Saturn's rings mostly bear alphabetical letters, in a scheme that is partly based on their distance from the planet, but also affected by their order of discovery. The A ring is the largest, easily seen from Earth with an outer edge 136,800 km (85,000 miles) from the centre of Saturn; the equally bright B ring lies just inside it, with an inner edge about 92,000 km (57,100 miles) from the centre. The rings are separated by the Cassini Division, an apparent 'gap' in the system some 4,700 km (2,920 miles) wide, while the narrower Encke Division subdivides the A ring itself. Such gaps are not entirely empty, but the thinner material orbiting within them appears much darker due to the contrast with its surroundings.

The A and B rings are bright because they are dominated by house-sized chunks of highly reflective water ice. Astronomers still aren't entirely certain where these particles came from, but one plausible theory is that they were created when a large outer moon, spiralling in towards Saturn, was broken up by tidal forces and stripped of its outer layers.

C and D rings

Orbiting closer to Saturn than the bright A and B rings lie two much fainter structures. The C or 'crepe' ring directly adjoins the B ring on its inner edge, while the tenuous D ring stretches down to a few thousand kilometres above Saturn's cloud tops. Particles in these two rings are much smaller and more scattered than those in their outer neighbours, and both rings are remarkably thin (as little as 5 m/16.4 ft deep on average), rendering them partially transparent.

Exactly why different rings have different-sized particles is still uncertain – one theory is that collisions between ice boulders in the brighter rings create grains of debris that slowly spiral inwards and, ultimately, sift down into Saturn's upper atmosphere. Certainly, structure in this region evolves over time – the Cassini probe tracked the motion of wavelike 'corrugations' at 30-km (19-mile) intervals across the inner rings, a pattern that may have been created by an injection of debris from a disrupted comet in the early 1980s.

The F ring and Prometheus

The fine structure of Saturn's rings is defined by a huge variety of gravitational influences. While the giant planet's gravity is dominant, the gravitational tug of war created by the influence of Saturn's many moons and countless 'moonlets' orbiting among the rings produces distinctive gaps and ringlets. In particular, gaps are found where the orbital periods of particles would 'resonate' with those of certain moons, in a similar process to the formation of the asteroid belt's Kirkwood gaps (see page 162).

The effect is at its most obvious in the F ring – a sharply defined ring just a few tens of kilometres wide that orbits just beyond the A ring. Here, particles are 'shepherded' into a single long-lived structure by the influence of the moon Prometheus orbiting on its inner edge. However, particles in the ring are subjected to constantly changing gravity from more distant moons, creating short-lived ripples and wavelike patterns. Prometheus is one of several known 'shepherd' moons that help maintain structure in both Saturn's ring system and those of other planets.

Outer rings

Beyond the F ring, Saturn's rings are looser and more diffuse. The first of these outer structures, known as the E ring, was discovered around the orbit of Enceladus in the 1960s, but many more have been found by space probes. Incomplete 'ring arcs' close to the planet often follow the orbits of very small moons and comprise microscopic particles of dust and ice blasted from the surface of the moons by 'micrometeoroid' impacts. The E ring, in contrast, is thought to be linked to geyser activity on Enceladus (see page 258). The curious G ring has a bright core (a cloud of metre-scale boulders centred on the small moonlet Aegeon), coupled with a dusty arc to complete the rest of the ring.

In 2009, astronomers using NASA's infrared Spitzer Space Telescope discovered the largest ring of all, a huge dust cloud produced by impacts on Saturn's outer moon Phoebe. Tilted at 27° to the other rings, it extends up to 12.4 million km (7.7 million miles) from the planet itself, and may be linked to the curious features of Phoebe's inner neighbour, Iapetus (see page 278).

Epimetheus and Janus

Lying just beyond the major rings, Epimetheus and Janus are a unique pair of moons that effectively 'timeshare' their orbits around Saturn. Heavily cratered and oval in shape, Janus has an average diameter of 179 km (111 miles), while the 145-km-diameter (90 mile) Epimetheus is more sharply chiselled – a clue that both worlds may be the broken remnants of a larger predecessor. The moons were both spotted for the first time in 1966, but were at first mistaken for a single object – an error that was not corrected until 1978.

The two moons coexist by a unique orbital dance – at times, Janus orbits about 50 km (31 miles) closer to Saturn than Epimetheus, but this means that it completes its orbit slightly faster. Every four years, as the two moons close in on each other, their mutual gravity slows the inner moon down and pulls it outwards, while accelerating the outer moon and causing it to drift inwards. Eventually, the two swap orbits and begin to move apart.

A Cassini image shows Epimetheus against the larger and more distant Titan, above Saturn's rings.

Mimas

The moons of Saturn grow larger out to the orbit of Titan, then smaller once again. The first of the 'mid-sized' moons (with orbits of a few hundred kilometres), is Mimas. At just 400 km (250 miles) across, it is the smallest known world in the solar system with sufficient gravity to pull itself into a sphere. This is largely due to composition – since Mimas is made mostly of ice, it is far easier to pull into shape than a rockier world.

Mimas is covered in craters; the largest, called Herschel, is some 130 km (80 miles) wide and 10 km (6 miles) deep. Herschel dominates an entire hemisphere, while on the moon's opposite side, huge fractures have formed in line with the crater rim. Yet despite a location directly in the firing line for objects drawn in by Saturn's gravity, Mimas's surface is not entirely saturated with craters. This suggests the moon was once resurfaced by a form of icy geological activity known as 'cryovolcanism'– eruptions of ice or water (perhaps mixed with ammonia to lower its freezing point) that wiped away many of the earliest craters.

Enceladus

Beyond Mimas, and past a handful of recently discovered small moonlets, Saturn's next mid-sized satellite is the intriguing Enceladus. Slightly misshapen, but with an average diameter of 504 km (313 miles), Enceladus has a brilliant-white surface (the most reflective in the solar system) and far fewer craters than might be expected.

Enceladus's landscape is blanketed in snow, created as water bursts in geyser-like eruptions from just beneath the surface, and freezes on exposure to the cold of space. Some areas are distinctly more cratered than others, suggesting that snows have fallen on different parts of Enceladus at different times in its history, while smooth plains may be a result of widespread cryovolcanic eruptions overwriting earlier terrain. Enhanced-colour images, meanwhile, reveal distinct variations in colour across the landscape, the most obvious of which are the bluish 'tiger stripes' near the moon's south pole, associated with the geyser outbursts.

Geysers of Enceladus

The geysers of Enceladus, like those on other planets, form when a reservoir of liquid heated to above its natural boiling point escapes through cracks in the surface, boiling suddenly and violently. In this case, the geysers' eruptions are powerful enough to fling material free of the moon's weak gravity, where it fuels Saturn's tenuous E ring. When NASA's Cassini probe flew through a geyser 'plume' in 2008, it discovered that it was mostly composed of pure water (with some traces of organic chemicals), escaping from a global ocean layer.

Enceladus's activity is thought to be driven by the same tidal heating seen among the satellites of Jupiter; in this case, Enceladus is caught in a gravitational tug-of-war between Saturn and the outer moon Dione. Currently, geyser activity is concentrated around the tiger stripes of the southern hemisphere, but it may migrate over time, and there are still some puzzles – principally, Enceladus seems to be warmer than the Saturn–Dione mechanism alone can explain.

Tethys

With a diameter of 1,062 km (660 miles), Tethys represents a step up in size from Saturn's innermost moons, and is a near-twin of Dione, the next moon out. Made almost entirely from ice, its density is actually less than that of water.

Tethys's surface is brilliant white, particularly on the leading hemisphere that faces 'forward' along its orbit. Here, the landscape is covered in a thin layer of 'snow' swept up from the outer edge of the E Ring (itself fuelled by the geysers of Enceladus). A broad swathe of terrain with fewer craters and a slightly darker colouring suggests that early in its history, icy eruptions wiped away some of Tethys's older craters. One hemisphere is dominated by the 400-km (250-mile) Odysseus crater. This vast but shallow impact basin has flattened over time as its ice has slumped back towards the average surface level at glacial speed. Elsewhere, faults and cracks in the surface show where the crust expanded as the interior froze. The longest of these is the 2,000-km (1,250-mile) Ithaca Chasma.

Telesto and Calypso

Remarkably, Saturn's mid-sized satellite Tethys shares its orbit with two other, much smaller worlds. Telesto and Calypso orbit Saturn precisely 60 degrees in front of and behind Tethys, respectively (in a rare example of a truly circular orbit). These positions put them at two of the Saturn–Tethys system's Lagrangian points, where the gravitational influence of the two larger worlds balances in such a way as to permit a stable orbit. Quite how these moons ended up in their current orbit is uncertain – they may be fragments of a larger satellite that broke up and were pushed towards their present locations by tidal forces, or they might even have formed alongside Tethys.

Physically, both moons resemble small misshapen asteroids, with average diameters of 25 km (15 miles) for Telesto (shown opposite) and 21 km (13 miles) for the slightly more elongated Calypso. Their surfaces are bright and unusually smooth, with traces of all but the largest craters smothered in a blanket of icy dust that they accumulate from Saturn's E ring.

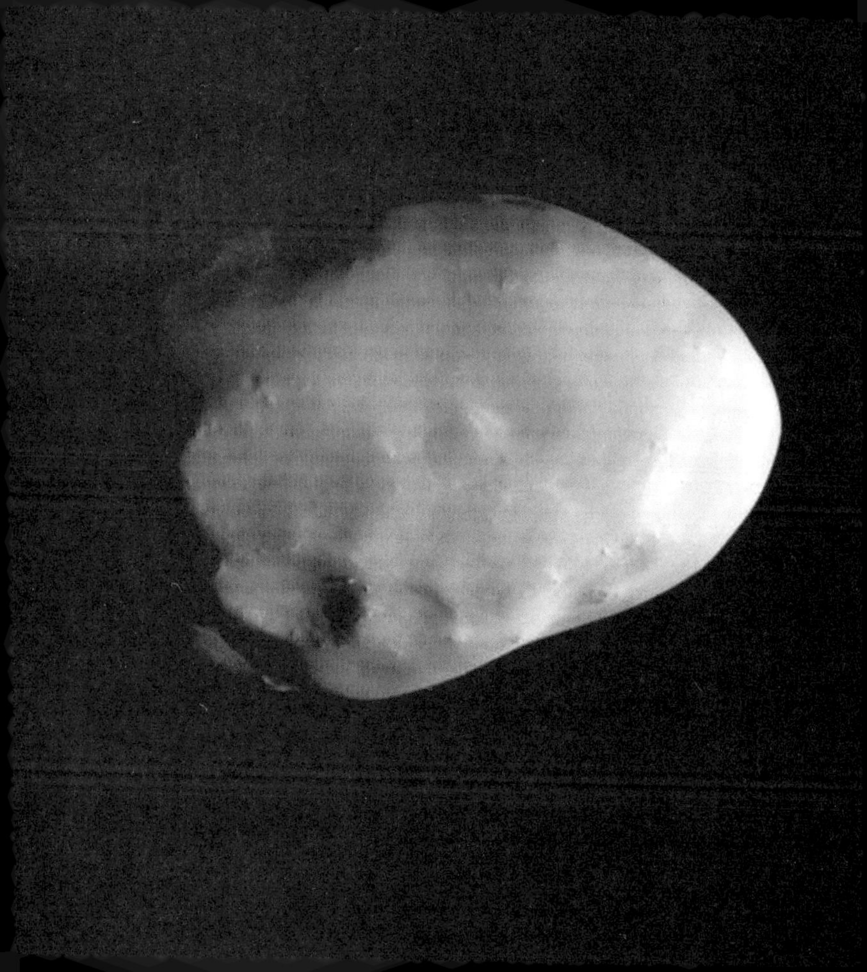

Dione

With a diameter of 1,123 km (698 miles), Saturn's moon Dione seems an obvious twin to its inner neighbour Tethys. It even has a similar pair of 'co-orbital' satellites, known as Helene and Polydeuces. However, while Tethys and Dione appear similar at first glance, closer inspection reveals some important differences. Space-probe flybys have confirmed that Dione is denser and considerably rockier than Tethys (although it still contains substantial ice). This may have allowed Dione to remain warmer and geologically active for longer, since it shows signs of more sustained icy cryovolcanism.

Like Tethys, Dione shows a history of heavy cratering, especially on its leading hemisphere. The trailing hemisphere, however, is criss-crossed by bright streaks of so-called 'wispy terrain'. In reality, these are towering cliffs up to several hundred metres high, with a reflective face of brilliant ice. They show that, at some point in its history, Dione endured huge tidal stresses that warped and broke its crust.

Rhea

Saturn's second-largest satellite, with a diameter of 1,528 km (949 miles), Rhea bears a striking resemblance to Dione. It, too, has a heavily cratered forward-facing surface and a darker trailing hemisphere criss-crossed by bright wispy terrain created by geological faulting. Craters on Rhea, however, are rather better defined than those on either Dione or Tethys, suggesting that its ice is less prone to slumping over time, perhaps because it is compressed to a greater density. Crater sizes vary between different parts of the moon, with some areas showing a mix of sizes and others only smaller impacts – evidence that Rhea was not always so deeply frozen, and that ice once erupted to wipe away some of the larger, earlier craters.

Scientists analysing space-probe images also identified another curious feature – a faint line of brighter material dotted around the planet's equator. One possible explanation is that Rhea once had a tenuous ring system, whose particles have since fallen to the surface, but the theory remains unproven.

Titan

The largest satellite in the Saturnian system, Titan has a diameter larger than the planet Mercury, and dwarfs all of its neighbours. It circles Saturn every 15.9 days, and spins on its axis in the same period.

Titan's strong gravity, coupled with the cold conditions some 10 AU from the Sun, allows the moon to hold onto a substantial atmosphere (something that the more massive Ganymede cannot do because of its warmer surface). As a result, Titan is shrouded in a dense orange haze that conceals its surface. It was only when the Cassini space probe arrived in the mid-2000s that near-infrared cameras finally pierced the atmosphere to reveal an eerily Earth-like world beneath. Titan's landscape mixes elevated continent-like regions with low-lying basins, river- and lakebeds and very few craters. While the moon appears to be made of a similar rock-ice mix to Saturn's other satellites, it clearly has a complex layered structure that gives rise to a variety of surface activity.

The atmosphere of Titan

Titan's atmosphere, like Earth's, is dominated by the unreactive gas nitrogen. However, a small amount of methane (CH_4), amounting to just 1.4 per cent by volume, renders the atmosphere opaque by producing clouds and a high-altitude haze that leaves the surface in a permanent orange twilight. Overall, the dense atmosphere exerts a pressure 1.45 times greater than Earth's own air at the satellite's surface.

The source of Titan's methane is something of a mystery, since unstable molecules in the gas break down when bombarded by solar radiation. Even at this distance from the Sun, the effect should be enough to remove Titan's methane completely within about 50 million years, so the gas must be replenished. As with the methane on Mars (see page 154), there are two plausible sources familiar from Earth – volcanic activity (most likely ongoing low-temperature cryovolcanism) or methane-producing microorganisms. A third possibility is that fresh methane somehow 'seeps' into the atmosphere from Titan's icy interior.

This Cassini image captures complex haze layers in Titan's upper atmosphere.

Titan's surface

The first close-up views of Titan's surface came from the European Space Agency's Huygens lander, which was released from the Cassini probe on its arrival at Saturn and parachuted into the moon's atmosphere in January 2005.

Huygens landed just off the 'coast' of one of Titan's elevated land masses. The probe was designed to float, in case it found itself landing in a liquid ocean, but in fact the landing site turned out to be a wide, flat plain littered with pebbles between 5 and 15 cm (2 and 6 in) wide. The underlying ground had the consistency of an ice-covered snowfield, but both soil and pebbles were tinted various shades of brown, most likely by hydrocarbon chemicals coating underlying water ice. The surface temperature was –179 °C (–290 °F), there was a gentle wind, and although there was no rainfall during the brief period that Huygens was able to send back data, the ground was damp – strong evidence to support the idea of a 'methane cycle' dominating Titan's weather.

A wide-angle view of Titan's surface, captured during the descent of the Huygens lander

Titan's methane cycle

Methane on Titan is thought to play a similar role to that of water on Earth. Conditions on the icy moon allow this simple hydrocarbon (chemical formula CH_4) to exist in liquid, solid and gaseous forms. Methane frosts coat deep-frozen water ice in the surface soil and rocks (alongside a mix of other hydrocarbons), while atmospheric methane condenses to form clouds, which precipitate methane rain and snow back to the surface, gradually eroding and softening the landscape as they run downhill.

Although scientists were somewhat disappointed when Cassini and Huygens data showed that Titan lacked widespread oceans, radar mapping subsequently revealed substantial lakes of liquid methane around the moon's north pole. Subsequent studies have suggested that the methane circulates in a climate cycle as complex as anything on Earth, accumulating in polar lakes during a particular hemisphere's seven-year 'autumn', returning to the atmosphere in 'spring', and being transported to the opposite hemisphere on prevailing winds.

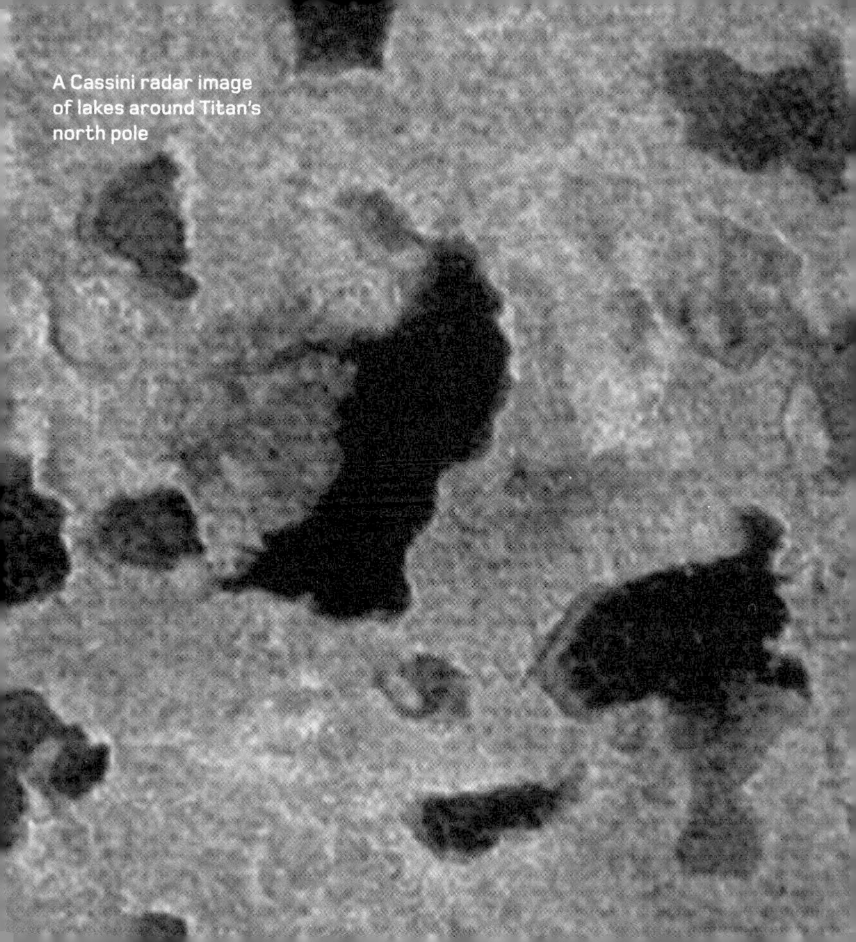

A Cassini radar image of lakes around Titan's north pole

Hyperion

Orbiting Saturn some way beyond Titan, Hyperion is a misshapen moon some 360 km long and 266 km wide (224 × 165 miles). This oval shape is unusual for such a large body whose gravity should have pulled it into a spherical shape, so many astronomers suspect that Hyperion is just a large surviving fragment of a much larger moon that was broken up in an ancient collision. The unusual shape may also be related to Hyperion's curious 'chaotic' rotation – it spins in unpredictable directions and at varying rates, in contrast to the orderly synchronous spin of neighbouring moons.

Hyperion's appearance is also bizarre – a spongelike structure of bright, razor-sharp ridges surrounding dark-floored pits. This strange landscape is a result of slow erosion, where the weak heat of the Sun causes ice to sublimate from darker patches of the surface, leaving the surviving rocky component to crumble inwards slowly. This unique process may only be possible because the moon's ancient interior has been exposed at the surface.

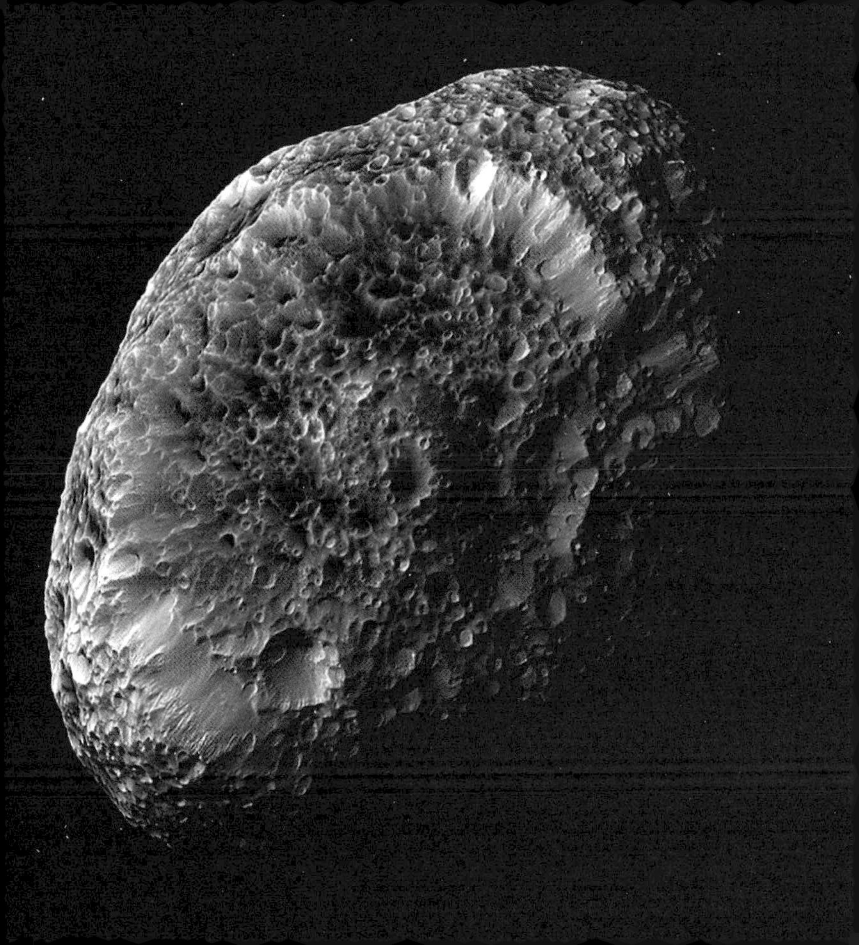

Iapetus

Saturn's third-largest moon, Iapetus is a world unlike any other in the solar system. Following its discovery in 1671, astronomers were puzzled by marked changes in brightness when seen at different positions in its orbit. The first space-probe flybys confirmed the long-held theory that the moon's two hemispheres have very different colours; the satellite's 'leading' face, which points along its orbit, is as black as coal, while its backward-facing or 'trailing' face is as bright as snow. Cassini images of Iapetus's surface showed the boundary between light and dark terrains in much more detail. The dark surface seems to overlay the brighter one, with no shades of grey between them. One popular explanation suggests that the dark material is rocky 'lag' left behind by the slow evaporation of ice from parts of a generally bright surface. Since dark surfaces absorb more heat than bright ones, the process would 'snowball' once it was begun, but how did it start? The best current theory is that the leading hemisphere picks up a thin film of dark dust from the Phoebe Ring (see page 250) as it moves along its orbit.

The equatorial ridge

Iapetus's starkly contrasting hemispheres are not its only curious feature – images from the Cassini space probe revealed a long, straight ridge that runs around much of the equator, giving the moon an overall shape similar to a walnut. This equatorial ridge is sharpest where it crosses a dark terrain called Cassini Regio. This central section is some 1,300 km (800 miles) long (about one-third of the moon's circumference), 20 km (12 miles) wide and up to 20 km (12 miles) high. In brighter regions, the ridge breaks down into smaller outcrops and mountains up to 10 km (6 miles) tall.

The origin of this remarkable feature is still hotly debated. One theory is that the ridge originated as an upwelling of ice from beneath the crust, and that tidal forces pulling on its extra mass slowly changed the moon's orientation to put it on the equator. Another is that the ridge is a remnant of a period when Iapetus rotated much faster and bulged out around its equator. A third alternative is that the ring is the remnant of a ring system that once orbited Iapetus and subsequently collapsed.

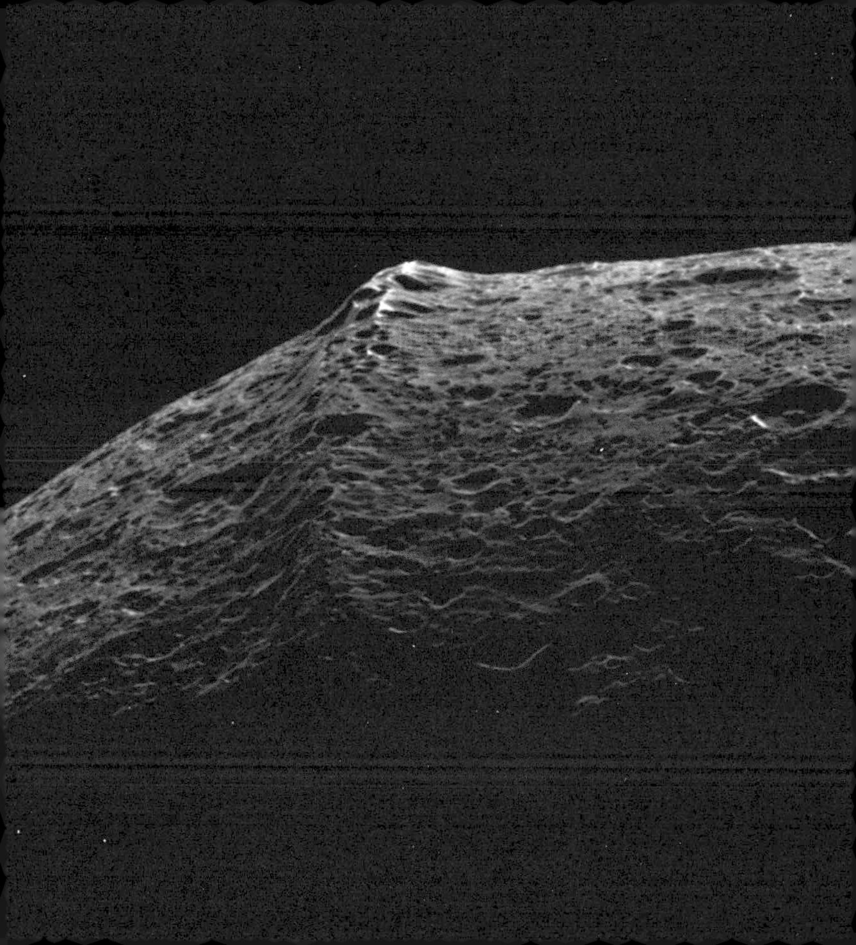

Phoebe

A large gap separates Iapetus, the outermost of Saturn's original satellites (which formed alongside the planet) from Phoebe, the largest and innermost of the many 'irregular' satellites that have been captured throughout the planet's long history. Phoebe's orbit, an average of almost 13 million km (8.1 million miles) from Saturn, is distinctly elliptical and also retrograde, orbiting the 'wrong way' around the planet.

When the Cassini probe flew past Phoebe in 2004, its images revealed an extremely dark, cratered world, roughly spherical and about 220 km (137 miles) across on average. Initial theories suggested that it was a captured asteroid, but relatively bright crater floors hint at the presence of ice just beneath the surface, and many astronomers now suspect the moon actually began life as a centaur (see page 26). Dark material that has been knocked off Phoebe's surface by small meteorite impacts now forms the huge outer Phoebe ring around Saturn, spiralling inwards where some of it is swept up by Iapetus.

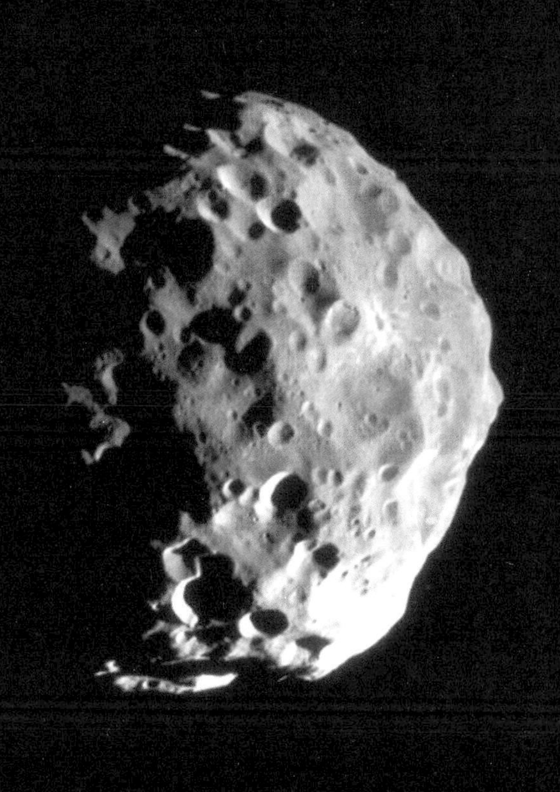

Moons of Saturn

Name	Diameter	Orbital period (days)	Eccentricity (circular = 0)
S/2009 S1	0.6 km (0.4 miles)	0.47	0
Pan	17x16x10 km (11x10x6 miles)	0.58	0
Daphnis	4.3x4.1x3.2 km (2.7x2.5x2.0 miles)	0.59	0
Atlas	20x18x9 km (12x11x6 miles)	0.60	0
Prometheus	68x40x30 km (42x25x19 miles)	0.61	0.00204
Pandora	52x41x32 km (32x25x20 miles)	0.63	0.0042
Epimetheus	65x57x53 km (40x35x33 miles)	0.69	0.009
Janus	102x93x76 km (63x58x47 miles)	0.69	0.007
Aegaeon	1 km (0.6 miles)	0.81	0.0002
Mimas	396 km (246 miles)	0.94	0.0202
Methone	3.2 km (2 miles)	1.01	0.0001
Anthe	2 km (1.2 miles)	1.04	0.001
Pallene	2.9x2.8x2.0 km (1.8x1.7x1.2 miles)	1.14	0.004
Enceladus	504 km (313 miles)	1.37	0.0045
Calypso	15x12x7 km (9x7x4 miles)	1.89	0.001
Telesto	16x12x10 km (10x7x6 miles)	1.89	0.001
Tethys	1062 km (660 miles)	1.89	0
Helene	22x19x13 km (14x12x8 miles)	2.74	0.005
Dione	1124 km (698 miles)	2.74	0.0022
Polydeuces	1.5x1.2x1.0 km (0.9x0.7x0.6 miles)	2.74	0.0192

Name	Diameter	Orbital period	Eccentricity
Rhea	1526 km (948 miles)	4.52	0.001
Titan	5150 km (3200 miles)	15.95	0.0292
Hyperion	180x133x103 km (112x83x64 miles)	21.28	0.1042
Iapetus	1469 km (913 miles)	79.33	0.0283

Name	Diameter	Period *	Ecc.
Kiviuq	~14 km (~8.7 mi)	449	0.334
Ijiraq	~10 km (~6.2 mi)	451	0.316
Paaliaq	~20 km (~12.4 mi)	687	0.364
Albiorix	~26 km (~16.2 mi)	783	0.469
Bebhionn	~6 km (~3.7 mi)	835	0.469
Erriapus	~8 km (~5 mi)	871	0.474
Tarqeq	~6 km (~3.7 mi)	888	0.16
Siarnaq	~32 km (~19.9 mi)	896	0.295
Tarvos	~14 km (~8.7 mi)	926	0.531
Narvi	~6 km (~3.7 mi)	1004 (R)	0.431
Bergelmir	~6 km (~3.7 mi)	1006 (R)	0.142
S/2006 S1	~6 km (~3.7 mi)	1015 (R)	0.13
Suttungr	~6 km (~3.7 mi)	1017 (R)	0.114
Hati	~6 km (~3.7 mi)	1039 (R)	0.372
S/2004	~6 km (~3.7 mi)	1046 (R)	0.401
Bestla	~6 km (~3.7 mi)	1084 (R)	0.521
Farbauti	~6 km (~3.7 mi)	1086 (R)	0.206
Thrymr	~6 km (~3.7 mi)	1094 (R)	0.47
Aegir	~6 km (~3.7 mi)	1117 (R)	0.252

* R = Retrograde orbit

Name	Diameter	Period *	Ecc.
S/2004 S07	~6 km (~3.7 mi)	1140 (R)	0.58
S/2006 S3	~6 km (~3.7 mi)	1227 (R)	0.471
Kari	~6 km (~3.7 mi)	1234 (R)	0.478
Fenrir	~4 km (~2.5 mi)	1260 (R)	0.136
Surtur	~6 km (~3.7 mi)	1298 (R)	0.451
Ymir	~18 km (~11.2 mi)	1312 (R)	0.335
Loge	~6 km (~3.7 mi)	1313 (R)	0.187
Fornjot	~6 km (~3.7 mi)	1491 (R)	0.206
Phoebe	109x109x102 km (68x68x63 mi)	548 (R)	0.164
Skathi	~6 km (~3.7 mi)	728 (R)	0.27
S/2007 S2	~6 km (~3.7 mi)	808 (R)	0.218
Skoll	~6 km (~3.7 mi)	878 (R)	0.464
Greip	~6 km (~3.7 mi)	921 (R)	0.326
Hyrrokkin	~8 km (~5 mi)	932 (R)	0.333
S/2004 S13	~6 km (~3.7 mi)	933 (R)	0.273
Mundilfari	~6 km (~3.7 mi)	953 (R)	0.21
Jarnsaxa	~6 km (~3.7 mi)	965 (R)	0.216
S/2007 S3	~6 km (~3.7 mi)	978 (R)	0.13
S/2004 S17	~4 km (~2.5 mi)	986 (R)	0.259

Uranus

Roughly twice as far from the Sun as Saturn, Uranus is the first of the solar system's two 'ice giant' worlds – a distinct class of planets intermediate in size between rocky worlds and gas giants, and dominated by chemicals with low melting points, such as water and ammonia. With a diameter of 50,724 km (31,518 miles), Uranus is four times larger than Earth but still less than half the size of its inner neighbour, Saturn.

The 1986 flyby of the Voyager 2 space probe revealed Uranus as a near-featureless turquoise ball, its distinctive colour created by a small amount of methane in the planet's outer atmosphere (2.3 per cent) that absorbs the red component of sunlight. However, studies with giant Earth-based telescopes have shown Uranus becoming far more active in the decades since Voyager's flyby, with obvious storm features erupting in its atmosphere. This suggests that the probe just happened to encounter Uranus during a particularly placid phase in its 84-year orbit around the Sun.

Inside the ice giants

Astronomers classify planets such as Uranus (opposite) and Neptune (shown in cross-section on page 319) as 'ice giants'. While the gas giants Jupiter and Saturn consist almost entirely of hydrogen and helium, in Uranus and Neptune these elements only dominate the outer atmosphere. About 5,000 km (3,100 miles) beneath the surface, they give way to a mantle dominated by chemicals such as water, ammonia and methane. These compounds (known in chemical terms as 'ices' because of their relatively low melting points) form a churning, slushy mantle around a rocky, roughly Earth-sized core. Changing temperature and pressure at different levels in the mantle can produce unusual chemical changes. For example, in Neptune at least, methane is thought to disintegrate at great depths, releasing heat and pure carbon that crystallizes into a 'rain' of diamonds. Swirling electric currents moving through the ices produce strong magnetic fields, but these are not symmetrical – instead (in both Uranus and Neptune), they are not only tilted sharply to the axis of rotation, but also offset from the planet's centre.

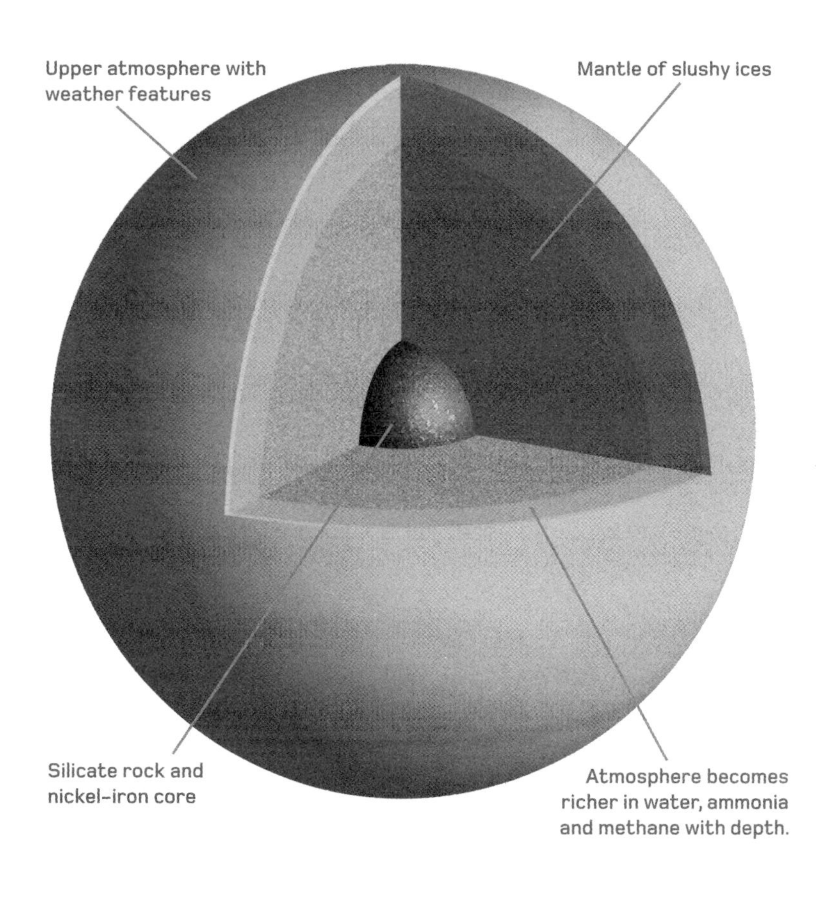

Upper atmosphere with weather features
Mantle of slushy ices
Silicate rock and nickel-iron core
Atmosphere becomes richer in water, ammonia and methane with depth.

Tilted planet

Uranus's most striking feature is its bizarre tilt, discovered in 1977 through studies of the orbits of its moons, and confirmed by the discovery of a ring system that makes the planet resemble a bullseye. While most planets have their axis of rotation tilted only moderately (if at all) from 'upright', Uranus's tips over at 98 degrees so that its north pole points slightly 'downwards' relative to its orbit. Astronomers suspect the planet's strange tilt is due to gravitational interactions with the other giants early in its history (see page 42). Today, however, it gives Uranus the strangest seasons of any planet. Regions close to the poles experience endless darkness througout a 40-year winter followed by 40 years of endless daylight in summer. Equatorial regions, in contrast, experience a day-night cycle more in tune with Uranus's rotation period of 17 hours 14 minutes. Around midwinter and midsummer, it's thought that the transport of heat around the planet, to balance out temperatures, suppresses the development of atmospheric features such as bands and storms.

A Hubble Space Telescope image of Uranus from 1999 reveals the planet's tilted orientation and bright surface storms.

The ring system and inner moons

Uranus's ring system is very different from the broad, bright planes seen around Saturn. Instead, the planet is encircled by 13 well-defined rings. The nine innermost of these are narrow but densely packed streams of orbiting particles; the next two are little more than dust trails; and the outer pair are diffuse and faint. Most of the ring particles have a distinctly reddish and unreflective coating, thought to be methane ice.

As in other ring systems, the streams of particles orbiting Uranus are influenced by nearby moons. The comparatively bright ε (epsilon) ring, for instance, is hemmed in by the influence of the shepherd moons Cordelia and Ophelia, while the outermost μ (mu) ring surrounds the orbit of another small moon, Mab. The rings show slight inclinations from the planet's equator, and along with the small but measurable eccentricity of the ε ring, these suggest the system is not entirely stable. Scientists are, therefore, fairly confident that the Uranian rings are less than 600 million years old.

An enhanced view of the inner rings as seen during the Voyager 2 flyby. The bright epsilon ring is on the right.

Miranda

Uranus's family of 27 known satellites is dominated by five larger worlds. With a diameter of just 472 km (293 miles), Miranda is the smallest of these, and also the closest to Uranus. Its surface is a mishmash of different terrains, formed by a variety of processes at different times in the moon's history. Verona Rupes in Miranda's southern hemisphere, for example, is one of the tallest cliff escarpments in the solar system, towering up to 5 km (3 miles) high. Elsewhere, there are racetrack-like structures of parallel grooves, ancient and heavily cratered plains, Ganymede-like sulci (see page 224), and even regions where icy eruptions seem to have obliterated earlier craters in the recent past. All of this suggests a long history of geological activity and cryovolcanism that may still be ongoing, but is hard to explain in terms of the 'tidal heating' effect found on moons such as Io. One possibility is that Miranda's orbit was once considerably more eccentric than it is now, driven by the influence of the outer moon Umbriel. This would have generated sufficient tidal heating to warm the moon's icy core, perhaps so much that it still hasn't entirely cooled down.

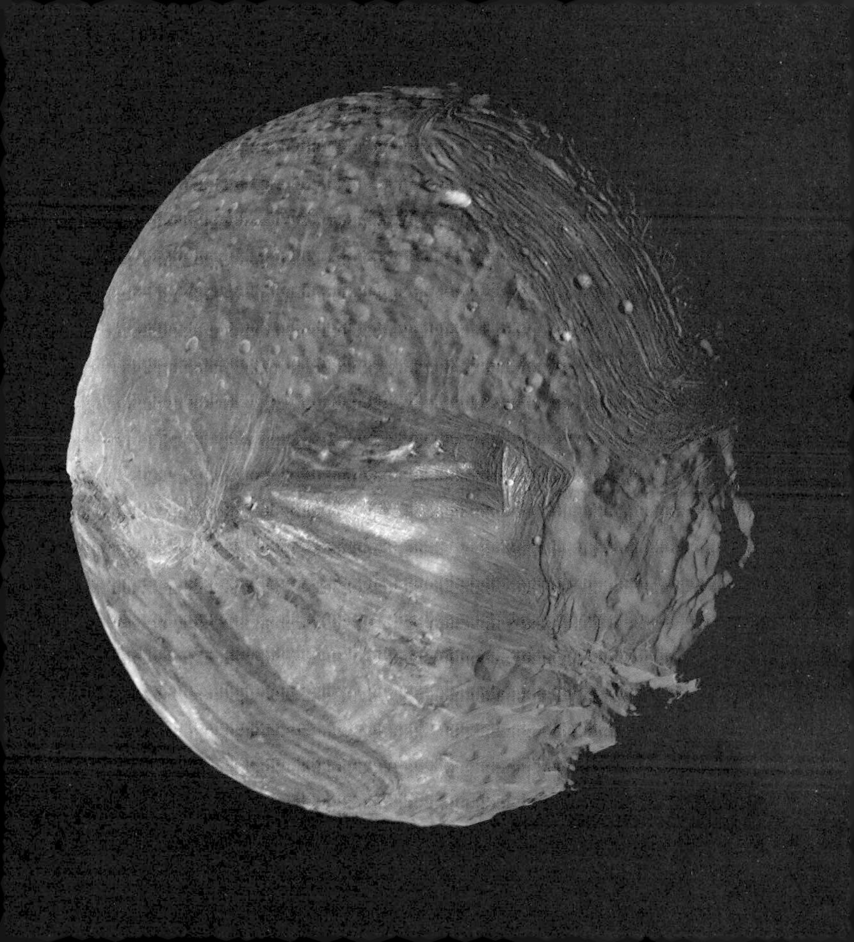

Ariel

The second major moon of Uranus, Ariel is considerably larger than Miranda, with a diameter of 1,158 km (720 miles). It has the brightest surface of any Uranian moon, generally grey in colour, but with a slight reddening on its leading hemisphere due to bombardment by particles trapped in the magnetosphere or by its parent planet. Based on estimates from crater counts, Ariel's surface is also the youngest of any Uranian moon, suggesting that cryovolcanic activity has resurfaced much of the moon with a fresh coating of ice since its early days. In places, deep canyons cut across the landscape – the largest, Kachina Chasmata, is 622 km (386 miles) long and up to 50 km (31 miles) wide. As with Miranda, Ariel's active history is hard to explain in terms of the tidal forces acting on the moon today, but computer models suggest that the satellite went through a period of orbital resonance with the larger Titania early in its history. After this produced significant tidal heating, Ariel's unusually high rock content may have been sufficient to trap heat that persists to the present day.

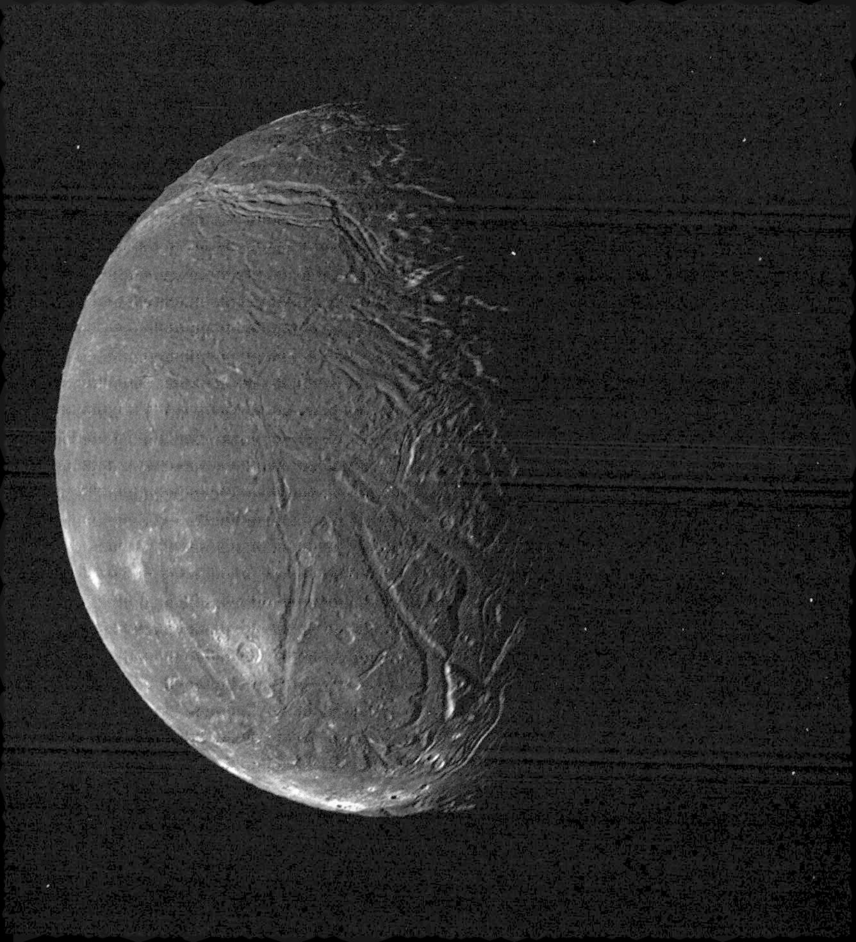

Umbriel

A near twin of Ariel in terms of size, Umbriel orbits somewhat further away from Uranus, circling the planet and spinning on its axis in 4.14 days. In contrast to bright Ariel, Umbriel is the darkest of the major satellites, with a slightly blue tint to its cratered surface. Broad canyons and variations in average brightness between different areas of the crust testify to activity shortly after the moon's formation, but the density of craters suggests that Umbriel has been geologically dead since the Late Heavy Bombardment. With an icier composition than Ariel, it was too small to retain much heat from its birth, while its orbit ensured it never experienced tidal heating.

A few large craters stand out from the rest due to their bright walls or central peaks; the most prominent of these, Wunda, is 131 km (81 miles) in diameter and sits on the moon's equator. The bright areas are probably freshly exposed water ice, but the absence of icy ejecta forming bright rays around these craters remains a mystery.

Titania

The largest Uranian satellite, Titania has a diameter of 1,578 km (981 miles) and resembles a bigger version of Ariel, with a relatively bright surface scored by deep canyons, and signs of widespread resurfacing once the worst of the early bombardments had subsided. Titania's density suggests it also has a similar composition to Ariel, with an equal mix of rock and ice. Combined size and rocky composition probably allowed Titania to retain heat from its formation and sustain geological activity, such as cryovolcanism, for some time (it orbits too far from Uranus to have ever experienced tidal heating). If the water ice is mixed with ammonia (as many think is possible given the chemical composition of the ice giants themselves), then its freezing point could be lowered considerably – perhaps enough for a subterranean ocean to survive today. Further questions surround the detection of carbon dioxide frosts alongside water ice on the surface – it's unclear whether the gas is produced by the action of solar radiation on surface minerals or whether it somehow escapes from reservoirs locked within the interior.

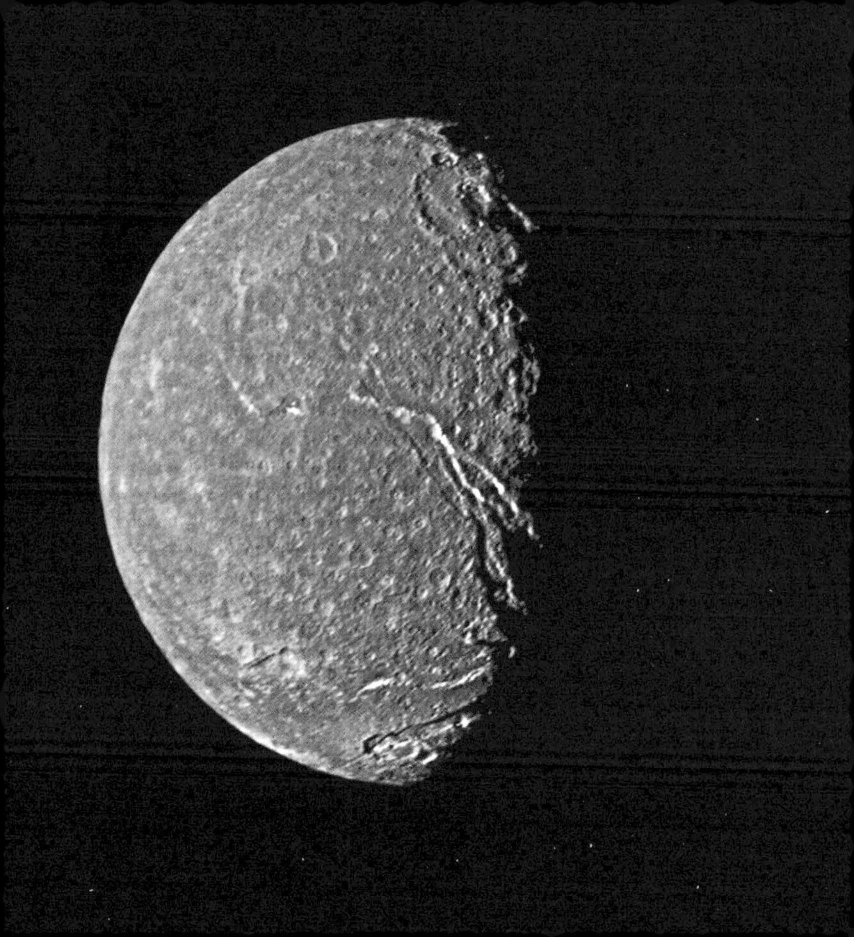

Oberon

The outermost major moon of Uranus, Oberon is barely 50 km (31 miles) smaller than Titania, and has a similar rock-ice composition. Images obtained by Voyager 2 suggest that it shares many of the features common among Uranian moons, including broad 'chasmata', heavy cratering and a roughly equal rock-ice composition. In terms of overall brightness, Oberon is the second-darkest moon after Umbriel, but its surface displays considerable variation. Bright rays, most likely formed as subsurface ice sprayed out during impact, blanket the landscape around many of the largest craters, but the crater floors are noticeably darker than other parts of the crust. The true nature of these dark patches is unknown – they may represent upwellings of material from cryovolcanic eruptions that followed major impacts or be due to the impacts themselves exposing a darker interior layer as they burrowed deeper. A prominent mountain seen on the moon's edge towers some 11 km (7 miles) high; it is probably the central peak of an unseen impact basin several hundred kilometres in diameter.

Moons of Uranus

Name	Diameter	Orbital period (days)	Eccentricity (circular = 0)
Cordelia	40 km (25 miles)	0.34	0.0003
Ophelia	42 km (26 miles)	0.38	0.0099
Bianca	54 km (34 miles)	0.43	0.0009
Cressida	82 km (51 miles)	0.46	0.0004
Desdemona	70 km (43 miles)	0.47	0.0001
Juliet	106 km (66 miles)	0.49	0.0007
Portia	140 km (87 miles)	0.51	0.0001
Rosalind	72 km (45 miles)	0.56	0.0001
Cupid	18 km (11 miles)	0.61	0.00007
Belinda	90 km (56 miles)	0.62	0.0001
Perdita	26 km (16 miles)	0.64	0.0012
Puck	162 km (101 miles)	0.76	0.0001
Mab	24 km (15 miles)	0.92	0.0025

Name	Diameter	Orbital period (days)*	Eccentricity (circular = 0)
Miranda	240x234x233 km (149x146x145 miles)	1.41	0.0013
Ariel	581x578x578 km (361x358x359 miles)	2.52	0.0012
Umbriel	1169 km (727 miles)	4.14	0.0039
Titania	1578 km (980 miles)	8.71	0.0011
Oberon	1523 km (946 miles)	13.46	0.0014
Francisco	22 km (14 miles)	267 (R)	0.1324
Caliban	72 km (45 miles)	580 (R)	0.1812
Stephano	32 km (20 miles)	677 (R)	0.2248
Trinculo	18 km (11 miles)	758 (R)	0.2194
Sycorax	150 km (93 miles)	1283 (R)	0.5219
Margaret	20 km (12 miles)	1695	0.6772
Prospero	50 km (31 miles)	1977 (R)	0.4445
Setebos	48 km (30 miles)	2235 (R)	0.5908
Ferdinand	20 km (12 miles)	2823 (R)	0.3993

* R = Retrograde orbit

Neptune

The solar system's outermost major planet, Neptune orbits the Sun at an average distance of 30.1 AU. Slightly smaller than Uranus, with a diameter of 49,244 km (30,600 miles), it is also a distinctly deeper shade of blue. This colour is in part due to the light-absorbing effects of methane, but since the proportion of the gas in each planet's atmosphere is the same, some other chemical must also play a role in generating Neptune's deeper colour.

Neptune's axial tilt of 28.3 degrees creates an Earth-like, though extremely elongated, cycle of seasons as the planet orbits the Sun every 164.8 years. In the absence of extreme temperature differences between the poles (thought to suppress activity on Uranus, see page 290), Neptune's 'normal' weather patterns were on clear display during Voyager 2's 1989 flyby. These range from deep, dark storms to fast-moving white clouds, formed by assorted chemicals condensing at different levels in the atmosphere.

Neptune's storms

Despite the feeble effects of heat from the Sun, Neptune is a surprisingly active planet, with violent weather powered in large part by heat escaping from the planet's interior. The slow contraction of slushy ices in Neptune's mantle triggers changes in the chemistry of compounds such as methane, releasing huge amounts of energy so that the planet radiates about 2.6 times more heat than it receives from the Sun.

Wind speeds of up to 600 metres per second (1,300 miles per hour) wrap narrow bands of wispy, high-altitude cloud around Neptune, and also propel compact storms of white cloud known as scooters. Larger storms take the form of dark oval spots – the so-called Great Dark Spot seen by Voyager 2 was originally assumed to be similar to Jupiter's long-lived Great Red Spot (see page 204). But these storms seem to last for just a few years at most, and instead of being high-altitude features, they actually seem to be clearings into the deeper, darker layers of Neptune's atmosphere.

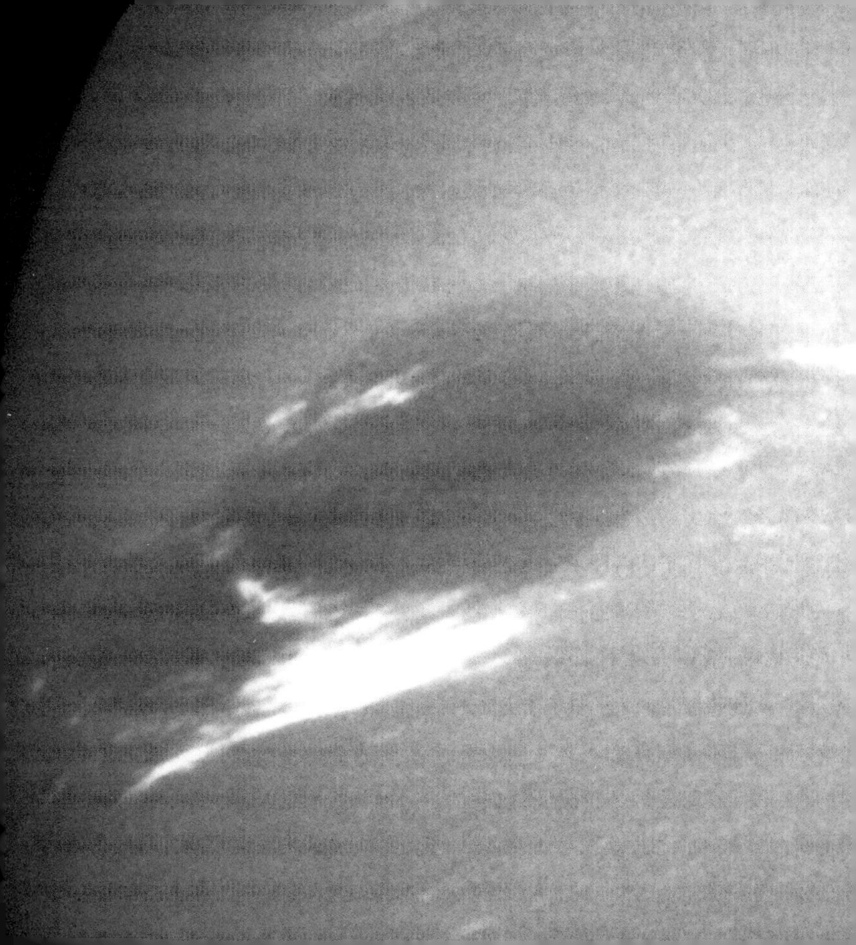

Rings of Neptune

Like all the giant planets, Neptune's gravity traps small particles in orbit around it, to form a system of rings. Narrow, dark and elusive, Neptune's rings have a reddish hue that hints at the presence of organic chemicals, such as methane ice. Early Earth-based attempts to spot occultations (dips in the light of distant stars) caused by intervening rings often produced contradictory results, and many scientists suspected that any rings might take the form of incomplete arcs. Voyager 2's flyby proved this was not too far from the truth – while Neptune's rings can be traced all the way around the planet, much of their material seems to clump together in certain regions, influenced by the gravity of shepherd moons. Five distinct rings are now recognized, each named after an astronomer who made an important early contribution to studies of Neptune. However, as with the rings of Uranus, the system seems to be relatively young and unstable – images taken in the decades since the Voyager flyby have shown a dramatic deterioration in their brightness and consistency.

A backlit image from Voyager 2 reveals the uneven brightness of Neptune's rings.

Triton

In contrast to the other giant planets, Neptune's satellite system is dominated by a single giant world that dwarfs 13 other known moons. Furthermore, the lone giant Triton (some 2,700 km or 1,680 miles wide) follows a retrograde orbit, circling the planet the 'wrong way'. This strongly suggests that it is a once-independent ice dwarf world that was captured into orbit around Neptune with catastrophic consequences for the planet's original satellites. With a surface temperature of around –235°C (–391°F), Triton is one of the coldest places in the solar system and we might expect it to be a deep-frozen, ancient world. In fact, its surface shows great variety, with few craters, a bright 'polar cap' at the south pole and a curious, pitted, 'cantaloupe terrain' covering much of one hemisphere. The 1989 Voyager 2 flyby even revealed active geysers spewing dust-laden nitrogen gas into a thin atmosphere. Triton's surprisingly active geology is probably linked to the events surrounding its capture – tidal heating that melted its interior allowed its heavier elements to form a core, from which heat is still escaping to this day.

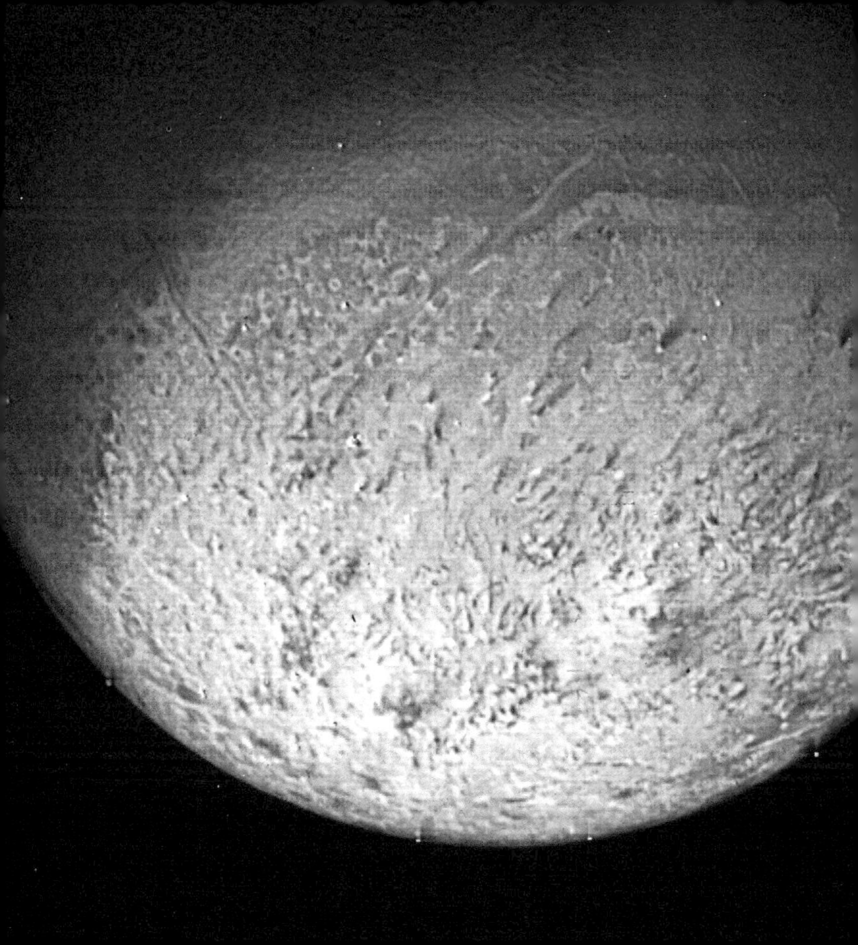

Proteus

Neptune's second-largest moon, Proteus has an average diameter of 420 km (261 miles), although its shape is distinctly elongated along an axis that points towards its parent planet. The moon's dark surface is pitted with countless craters, the largest of which, known as Pharos, is itself about 240 km (150 miles) across. Such a large impact should have been enough to shatter a solid world, and this fact, along with Proteus's shape, offers another clue to the moon's true nature.

Proteus is thought to be a second-generation moon that coalesced out of debris created in the chaos of Triton's arrival in the Neptunian system. Computer models suggest that even close to the planet, Neptune's original moons were unlikely to have survived intact (most likely falling victim to collisions with other satellites), but once the havoc subsided and Triton's orbit developed its current circular shape, new moons could form at a safe distance. Loosely structured and shaped by the tidal influence of Neptune itself, Proteus is the largest of these.

Nereid

The third-largest moon of Neptune (and the second to be discovered), Nereid follows an elliptical path that ranges between 1.37 million km and 9.64 million km (851,300 and 5.99 million miles) from Neptune. This kind of extreme orbit is typical of the outer solar system's 'irregular' moons (asteroids and comets captured into orbit around the giant planets). However, Nereid's diameter of roughly 340 km (211 miles) would be unusually large for such a captured world. Furthermore, the properties of sunlight reflected off its surface are distinctly different from those seen in centaurs (see page 338). In fact, Nereid's surface composition seems to be a closer match for Uranus's moons Titania and Umbriel. A smaller irregular satellite called Halimede, with a similar surface to Nereid, may be a fragment of the larger moon broken off during a collision. Coupled with the traumatic history of the Neptune system, all this evidence raises the intriguing possibility that Nereid is a surviving member of Neptune's original satellite family, ejected to its current orbit by the arrival of Triton.

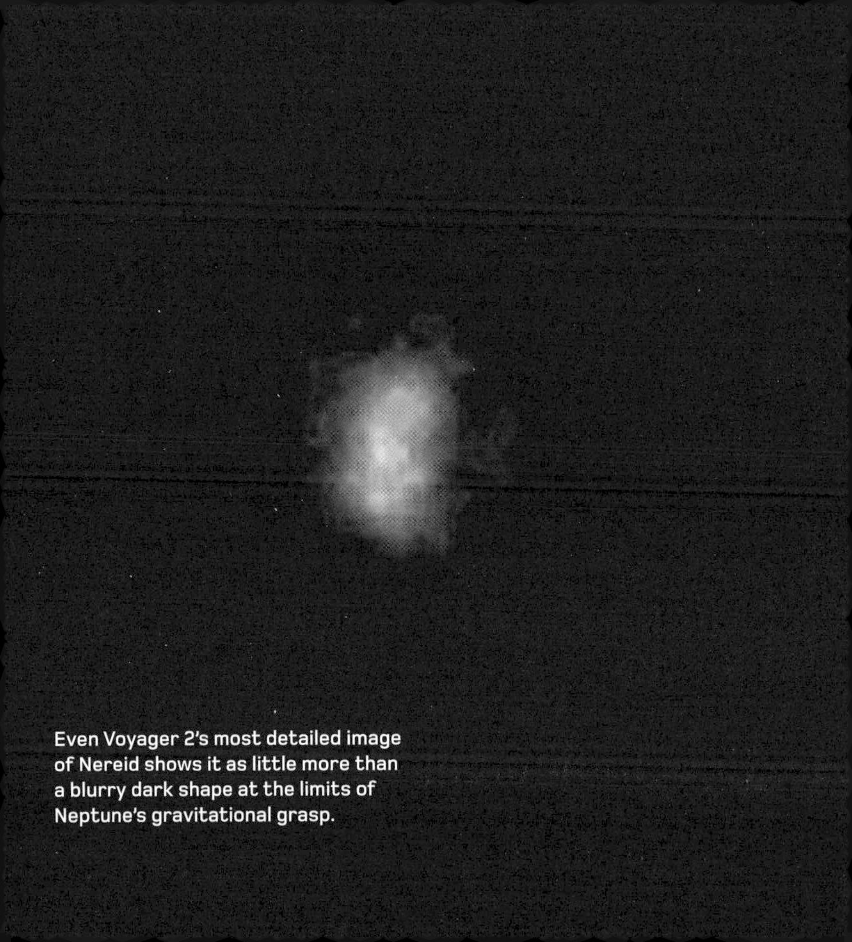

Even Voyager 2's most detailed image of Nereid shows it as little more than a blurry dark shape at the limits of Neptune's gravitational grasp.

Moons of Neptune

Name	Diameter	Orbital period (days)*	Eccentricity (circular = 0)
Naiad	48x30x26 km (30x19x16 miles)	0.29	0.0003
Thalassa	54x50x26 km (34x31x16 miles)	0.31	0.0002
Despina	90x74x64 km (56x46x40 miles)	0.33	0.0001
Galatea	102x92x72 km (63z57x45 miles)	0.43	0.0001
Larissa	108x102x84 km (67x63x52 miles)	0.55	0.0014
S/2004 N1	20 km (12 miles)	0.95	0
Proteus	220x208x202 km (137x129x126 miles)	1.12	0.0004
Triton	2707 km (1682 miles)	5.88 (R)	0.000016
Nereid	340 km (211 miles)	360	0.7512
Halimede	60 km (37 miles)	1880 (R)	0.571
Sao	40 km (25 miles)	2914	0.293
Laomedeia	40 km (25 miles)	2168	0.424
Psamathe	40 km (25 miles)	9116 (R)	0.45
Neso	60 km (37 miles)	9374 (R)	0.495

* R = Retrograde orbit

Interior of Neptune

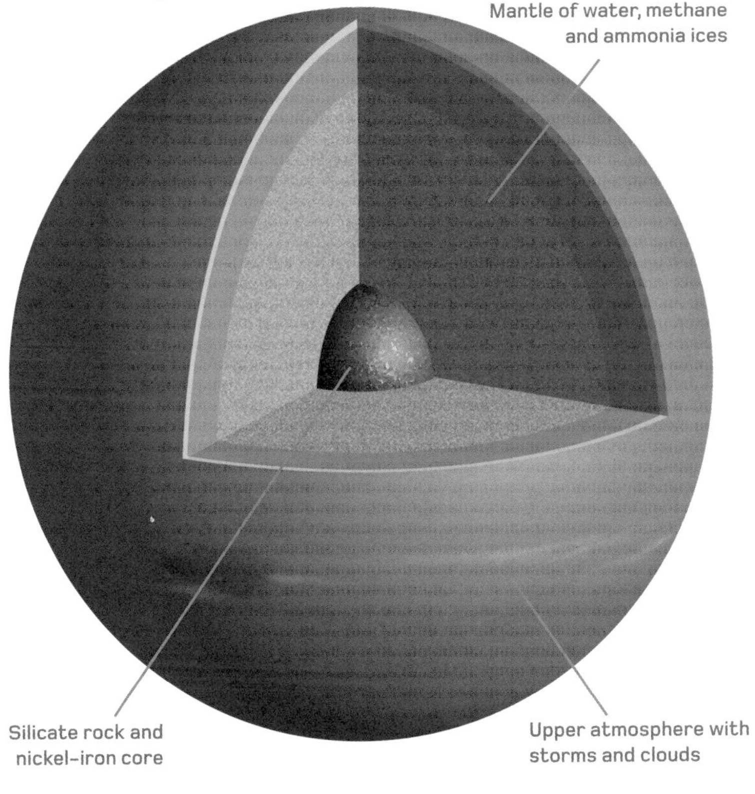

Mantle of water, methane and ammonia ices

Silicate rock and nickel-iron core

Upper atmosphere with storms and clouds

Comet orbits

The vast majority of comets orbiting the Sun reside far beyond the realm of the planets, following more or less circular paths within the enormous spherical halo of the Oort Cloud (see page 370). Those that enter the inner solar system, however, follow highly elongated ellipses that bring them close to the Sun at one end (perihelion), and are grouped into several families depending on their orbital characteristics. Those with orbits of less than 200 years (known as 'short-period' or 'periodic' comets) have their aphelion – the furthest point in their orbit – among the giant planets or just beyond in the Kuiper Belt. 'Long-period' comets reach aphelion much further out, perhaps returning to the Oort Cloud from whence they came. 'Hyperbolic' comets are flung out of the solar system completely after their single encounter with the Sun. The gravitational influence of the giant planets, particularly Jupiter, can transform a long-period comet into a short-period one, and even circularize its orbit, making it virtually indistinguishable from a centaur (see page 336) or asteroid.

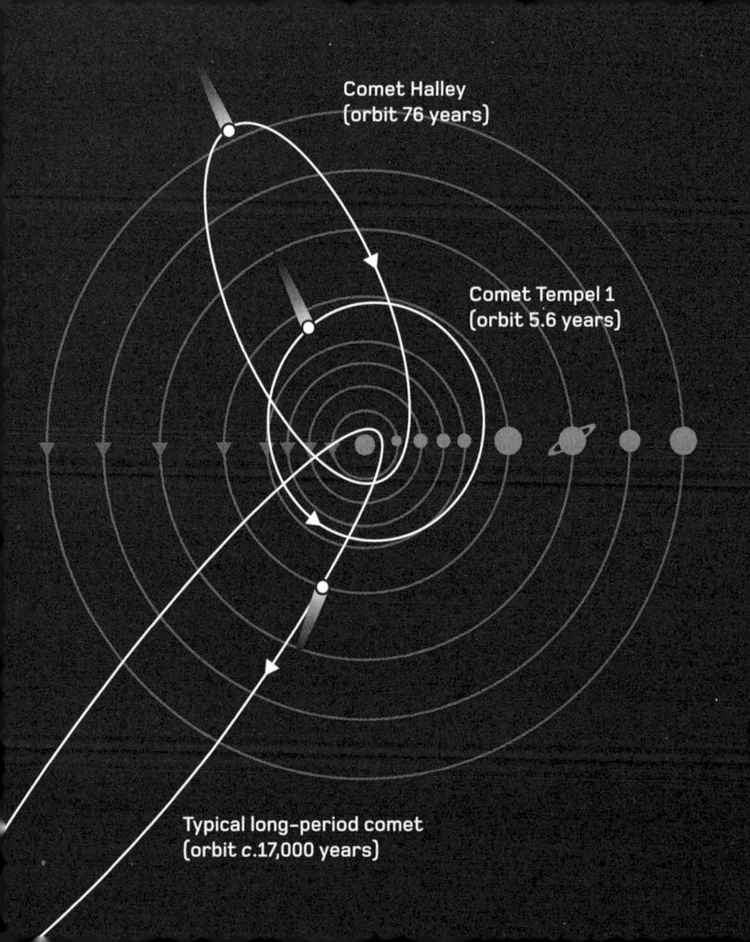
Comet Halley
(orbit 76 years)
Comet Tempel 1
(orbit 5.6 years)
Typical long-period comet
(orbit c.17,000 years)

Comet Ikeya-Seki

The most spectacular comet of recent decades, Comet Ikeya-Seki was discovered in September 1965, as it was already about to cross inside Earth's orbit bound for a rendezvous with the Sun. Like all comets, it spends most of its orbit in a dormant state, and only becomes active when, heated by the Sun, ice in its solid nucleus vaporizes to form a huge gas halo called a 'coma'. Escaping jets of vapour are caught up on the solar wind and blown away from the Sun to form an elongated tail. Ikeya-Seki's activity was particularly impressive – the comet rapidly grew bright enough to see in daylight, and its tail extended across more than 113 million km (70 million miles) of space. On 21 October, as the comet passed within 450,000 km (280,000 miles) of the Sun, it broke into three fragments that were tracked on their retreat into the outer solar system on orbits that will see them return in about 1,000 years. Curiously, astronomers think something like this happened to the comet once before – it is a member of the 'Kreutz sungrazers', a family of comets in similar orbits, thought to originate from the break-up of an earlier 'great comet' seen by skywatchers in 1106.

Comet Wild 2

This small and unremarkable comet, discovered in 1978, is one that we know most about, since parts of it have been brought back to Earth. Wild 2 was discovered on its first passage through the inner solar system, several years after a close encounter with Jupiter's gravity, which changed its orbit from a more or less circular one of 43 years' duration to a more elliptical track with a 6.4-year period. This brings it within 1.6 AU of the Sun at its closest approach, and made it an ideal target for NASA's Stardust mission, which visited the comet in 2004 and returned samples from its coma and tail in 2006. Laboratory analysis of Wild 2's dust suggests that the comet formed in cold conditions at a great distance from the Sun, but incorporated dust from hotter regions, blown outwards on the solar wind. Subsequently, it spent most of its life in the cold outer reaches of the solar system before its orbit was disrupted to bring it closer to the planets. The dust also contained a wide variety of organic (carbon-based) chemicals, supporting the idea that comets might kick-start the development of life by delivering such chemicals to newly formed planets.

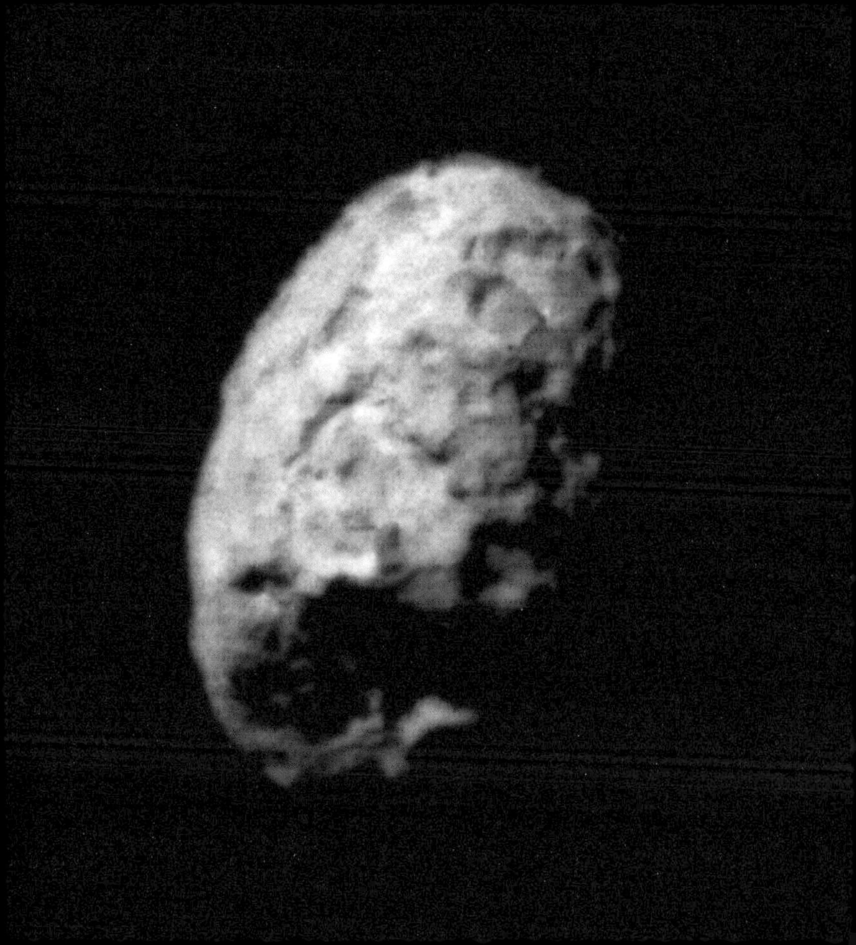

Comet Churyumov-Gerasimenko

Often known by its catalogue number as Comet 67P, this small icy world is thought to have originated in the Kuiper Belt, but currently follows a 6.4-year orbit that takes it out to the orbit of Jupiter at aphelion, and brings it within 1.25 AU of the Sun at perihelion. It was the destination for the European Space Agency's hugely successful Rosetta mission, which orbited 67P for two years from 2014 (throughout its perihelion passage) and placed a lander, called Philae, on its surface.

Structurally, Comet 67P consists of two distinct, irregularly shaped lobes, joined by a narrow 'neck'. The different orientations of exposed layers within the lobes suggest that they began life as separate objects and, subsequently, stuck together after a low-energy collision. Rosetta's surveillance revealed just how traumatic a comet's perihelion passage can be; as ice escaped from beneath the surface, the probe recorded collapsing cliffs, boulders flung across hundreds of metres and large pits opening in the landscape.

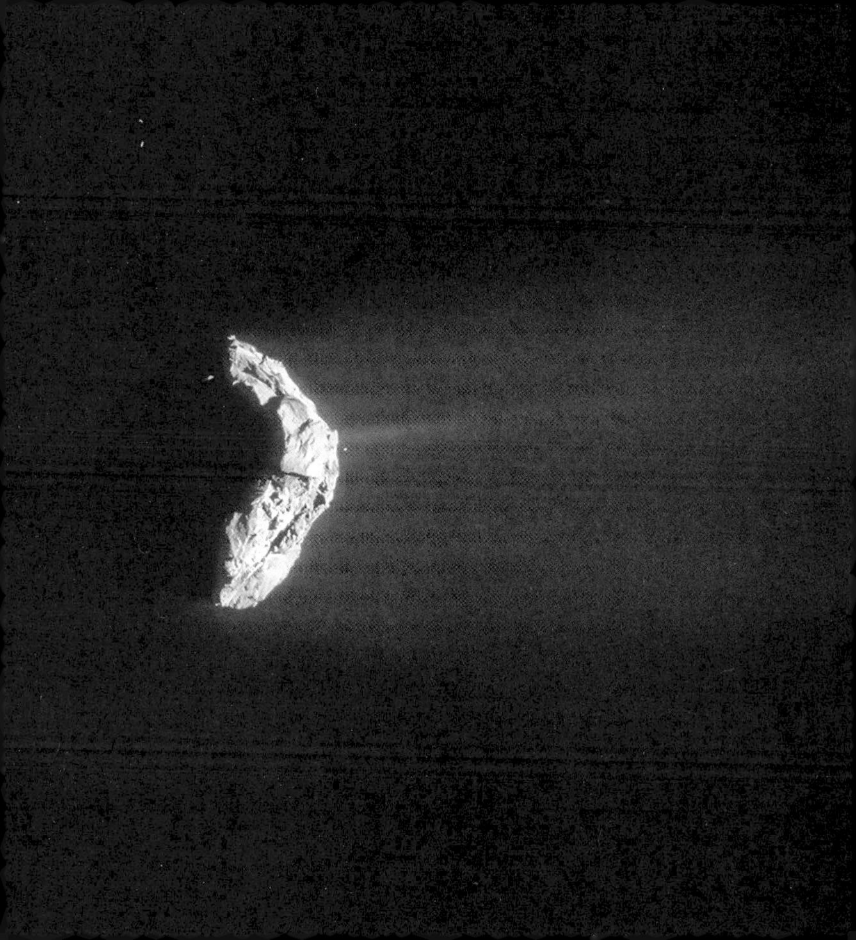

Comet Encke

This curious comet was the second short-period comet to be identified, more than a century after the far more famous Comet Halley (see page 340). Encke's orbital period of just 3.3 years is one of the shortest known, taking it from the outer asteroid belt to within the orbit of Mercury. Comets 'use up' their reserves of ice with each perihelion passage, yet Encke is still moderately active and so is thought to have arrived in its current orbit quite recently.

Encke leaves a trail of debris behind it that is thought to be the source of the 'Taurid' meteor showers seen when Earth comes near the comet's orbit in June/July and November. Close approaches between Earth and the comet itself happen every 33 years and, while it currently presents no danger, Encke's frequent encounters with the strong gravity of the rocky planets result in its orbit evolving over time. Some astronomers have even suggested that, in earlier times, Encke was much brighter and came closer to Earth, inspiring fear in our prehistoric ancestors.

Comet Shoemaker–Levy 9

Famous as the comet that smashed into Jupiter with spectacular results in July 1994 (see page 210), Shoemaker–Levy 9 was discovered a little over a year earlier, already in orbit around the giant planet. It is thought to have been captured by Jupiter during a close encounter some time around 1970. The comet survived its capture intact, initially following an orbit with a roughly two-year period.

A close encounter then brought it within 40,000 km (25,000 miles) of Jupiter's cloud tops in July 1992. This is closer even than the innermost Jovian moon Metis, and well inside Jupiter's 'Roche limit' – the region where the tidal forces exerted by the giant planet's gravity become powerful enough to overcome the forces holding other large objects together. As a result, the cometary nucleus disintegrated into a series of fragments up to 2 km (1.2 miles) across, strung out into a 'string of pearls' as they made their final approach to Jupiter. Crater chains found on the surfaces of Callisto and Ganymede seem to suggest this is a common fate for comets captured by Jupiter's gravity.

Comet Tempel 1

When first identified as a periodic comet in 1867, the faint short-period Tempel 1 followed a predictable 5.7-year orbit that took it from the outer asteroid belt to a perihelion just beyond Earth's orbit and back. Astronomers lost track of it in the late 19th century, however. It was presumed destroyed until 1967, when new research showed how it had been disrupted by close approaches to Jupiter in 1881 and, later, in the 1940s and 1950s. These interactions had first lengthened, then shortened the orbit – when subsequently rediscovered, it was found to have settled into a new 5.5-year pattern.

Tempel 1's 14-km (9-mile) nucleus made it an ideal target for NASA's Deep Impact probe, which encountered the comet in 2005, fired a barrel-like projectile into its surface and studied the fountain of material blasted into space by the impact. Analysis showed that Tempel 1's material seems to have gone through a number of chemical changes – it was not the pristine relic from the early solar system that most had expected.

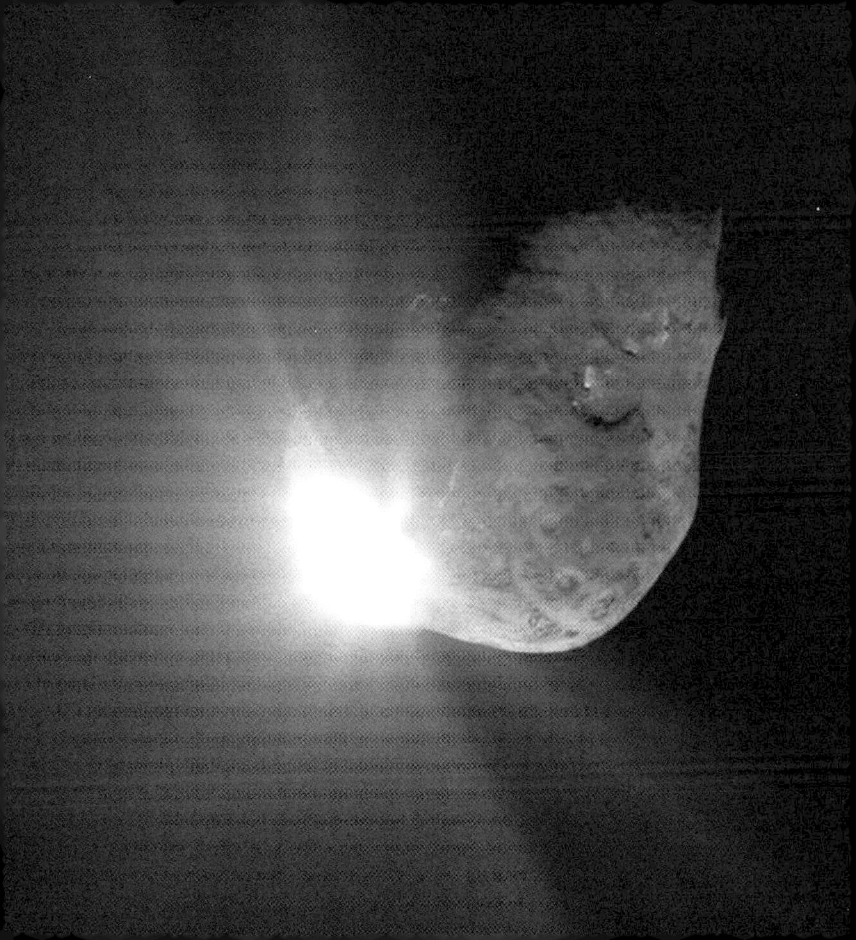

Hartley 2

After releasing its impactor probe at Comet Tempel 1, NASA's Deep Impact mission flew on to a rendezvous with another comet in November 2010. Roughly 1.1 km (0.7 miles) across and shaped rather like a bowling pin, Hartley 2 is the smallest comet so far to have been visited by a space probe. Orbiting the Sun in just under 6.5 years, it reaches aphelion beyond Jupiter, but comes with 0.05 AU of Earth's orbit at perihelion.

Studies of Hartley 2 offered a possible solution to one mystery about Earth's origins. While it was long assumed that much of our planet's present-day water originates from comets, close-up studies of comets themselves consistently found that comet ices contained too much 'heavy' water (made with deuterium, a heavier variant or isotope of hydrogen that is rare on Earth). Water vapour in Hartley 2's coma, however, seems to match exactly with the characteristics of water on Earth. The difference in water content may be related to evidence that this comet formed closer to the Sun and in warmer conditions than its longer-period relatives.

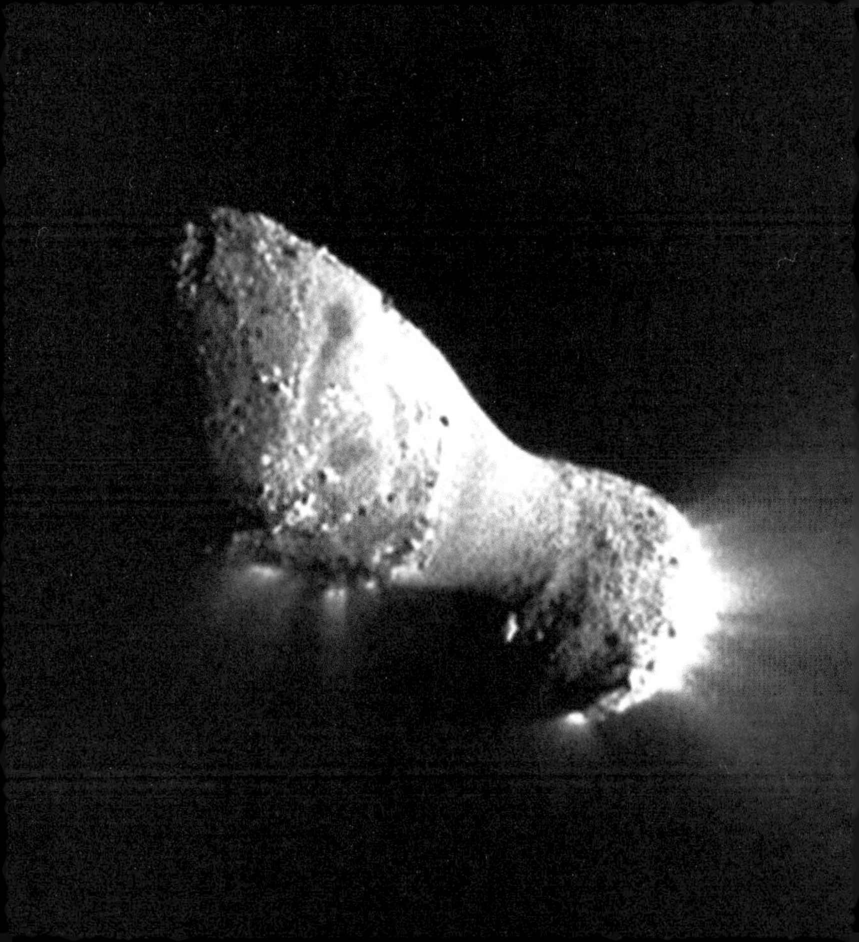

Chiron

Discovered in 1977, Chiron bears the telltale catalogue number of an asteroid, but was in fact the first object in an entirely new class – the centaurs (see page 26). With a diameter of roughly 218 km (135 miles) according to the latest estimates, Chiron spends most of its 50-year orbit between those of Saturn and Uranus, though it crosses both around perihelion and aphelion.

At first, Chiron was thought to be a refugee from the asteroid belt – it was even given the asteroidal designation 2060 Chiron. Then, in 1989, astronomers noticed that it was growing steadily brighter and new photographs revealed the presence of a cometlike coma. This earned it the cometary designation 95P/Chiron, making it a rare object classed as both a comet and an asteroid. Unfortunately, little is known about Chiron's physical properties, except that it has water ice on its surface, and that it is the largest in a subgroup of centaurs with dark, blue and grey surfaces (in contrast to Pholus, see page 338).

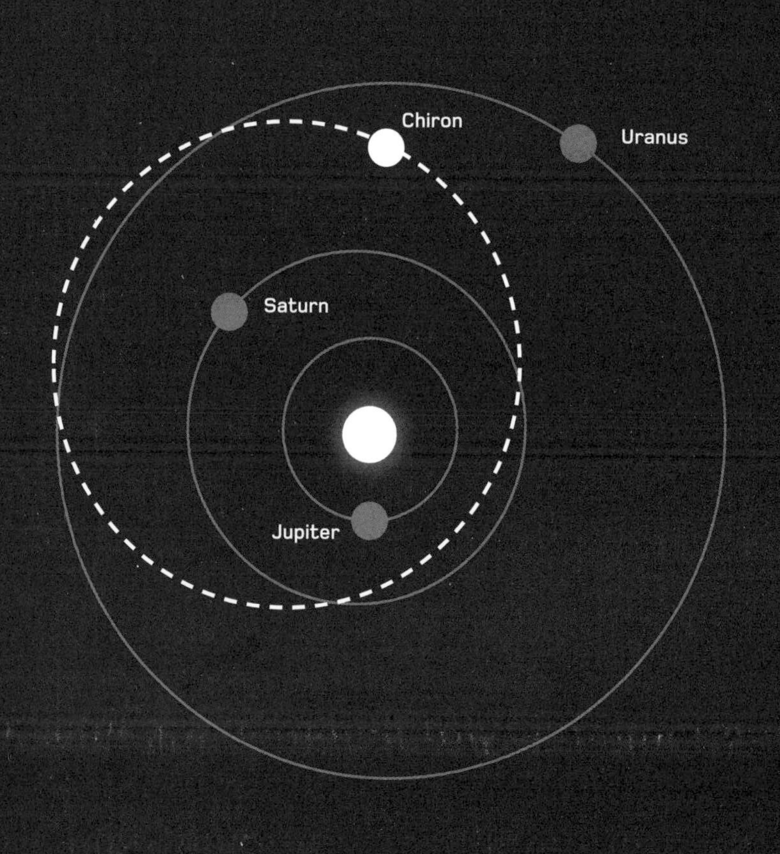
Chiron
Uranus
Saturn
Jupiter

Pholus

The discovery of Pholus in 1992 led to the realization that Chiron was not alone, and that there are a significant number of centaur objects orbiting among the giant planets. Pholus bears the asteroid number 5145, and has a 92-year orbit that is more elliptical than Chiron's, going from inside the orbit of Saturn to outside that of Neptune, and back.

Physically, Pholus is slightly smaller than Chiron at about 180 km (112 miles) across. Its surface is brighter and distinctly red in colour, making it the prototype for the second major centaur subgroup. Centaurs seem to be either dark and blueish or bright and reddish, with no obvious continuum between the two types. Most experts agree that the redder ones carry significant amounts of carbon-based 'organic' chemicals, reddened by exposure to radiation from the Sun. But attempts to divine the origin of the two categories have been frustrated by other shared characteristics; the red and blue centaurs mix in similar orbits and seem equally likely to show cometary activity.

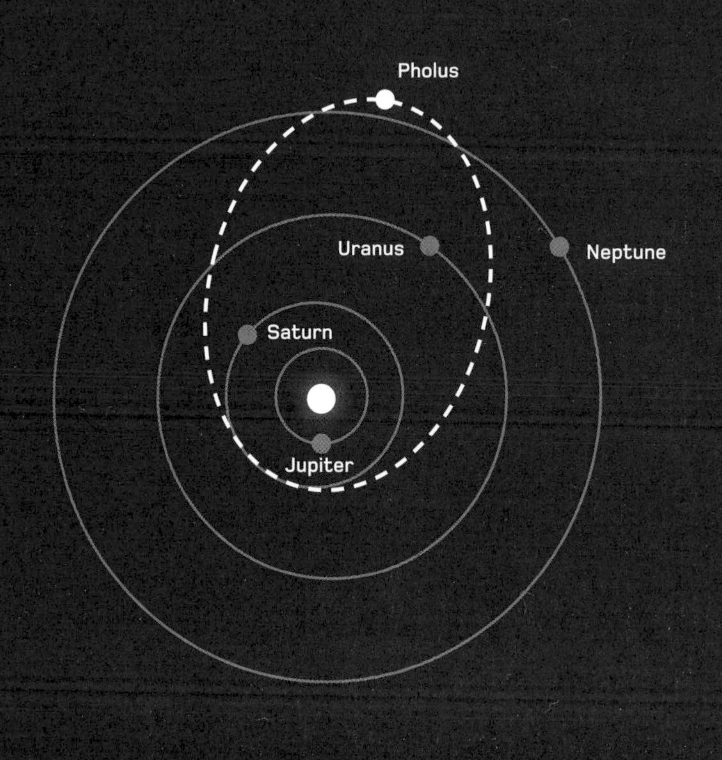
Pholus
Uranus
Neptune
Saturn
Jupiter

Comet Halley

The most famous comet of all, Halley was the first to have its periodic nature recognized. Edmond Halley realized that comets seen in 1531, 1607 and 1682 were, in fact, a single object, and predicted that it would return once again in 1658. Apparitions at more or less regular 76-year intervals have since been traced back to at least 240 BCE. The nature of its orbit suggests that Halley is one of the fairly uncommon family of comets that originate in the Oort Cloud but have their orbits shortened by interaction with the giant planets.

Halley's most recent perihelion passage occurred in 1986, and although the geometry did not favour observers on Earth, an armada of spacecraft were sent to greet it. The European Space Agency's Giotto probe flew within 600 km (373 miles) of the 16 × 8 km (10 × 5 mile) nucleus and returned images of a dark surface and violent jets. Dust blown off the nucleus but still following Halley's orbit gives rise to two regular meteor showers on Earth, known as the Eta Aquariids and the Orionids.

Comet Hale–Bopp

Thought to be the most widely observed comet of all time, Hale–Bopp was visible to skywatchers on Earth for 18 months. At its brightest, it outshone every star in the sky but one. Discovered in 1995 while it was still well beyond the orbit of Jupiter at a distance of 7.1 AU, early measurements suggested that it last visited the Sun 4,200 years ago. However, a close encounter with Jupiter in March 1996 saw its orbit change considerably and its next return will take a mere 2,380 years.

Hale–Bopp's early brightness indicated a substantial nucleus, estimated at 60 km (37 miles) wide, although it could not be observed directly. Distinct jets emerging from the surface could be identified within the coma, allowing the comet's rotation period to be calculated as 11 hours and 46 minutes. Some astronomers claimed the full pattern of Hale–Bopp's activity could only be explained by the presence of a substantial satellite nucleus orbiting the main one, but this intriguing theory could not be confirmed before the comet retreated into the depths of space.

Comet ISON

Comets are unpredictable by nature, and even those that promise to make brilliant apparitions as they round the Sun can ultimately disappoint. Comet ISON, for example, was expected to make a spectacular appearance as it passed Earth on its retreat from perihelion passage in November 2013, but never achieved its potential. The comet was discovered using a telescope from the International Scientific Optical Network (ISON) almost a year before perihelion, well outside the orbit of Jupiter, but already displaying a substantial coma. It approached on a hyperbolic trajectory – a path that would take it past the Sun just once before being flung into interstellar space – and was therefore either a fresh arrival from the Oort Cloud or, less likely, a visitor from beyond the solar system itself. However, as the comet passed within 1.17 million km (727,000 miles) of the Sun, it disintegrated. Most of the nucleus was destroyed, but a small fragment was later found, still continuing along its original trajectory with a greatly reduced brightness.

Pluto

The brightest, and still the largest-known Kuiper Belt Object (KBO), with a diameter of some 2,377 km (1,477 miles), Pluto was discovered in 1930, and was historically considered a planet in its own right. Today, it is simply the largest dwarf planet (see page 16).

Circling the Sun every 248 years, Pluto's elliptical orbit carries it between 29.7 and 49.3 AU from the Sun. At perihelion it is closer than Neptune, but its orbit is tilted at 17 degrees to the plane of the solar system and there is no risk of collision. As a result of its great distance and tiny size, it remained a mystery for decades. A giant moon, Charon (see page 350), was discovered in 1978, and four much smaller satellites have since been found. The orientation of their orbits reveals that Pluto's axis of rotation is 'tipped over' at an angle of 120 degrees, giving it a Uranus-like pattern of extreme seasons. Other Earth-based observations confirmed the presence of a sparse nitrogen atmosphere, and a surface covered with nitrogen ice and other frozen gases.

Pluto's surface

When New Horizons flew past Pluto in July 2015, it revealed an extraordinary world. Only one hemisphere could be mapped, since the planet's southern half was in seasonal darkness, but the northern side alone proved to contain a surprising variety of terrains. A heart-shaped bright region consists of two distinct lobes across which the density of cratering suggests two different ages. The western lobe, known as Sputnik Planitia, is particularly young, suggesting Pluto has seen cryovolcanic activity in its recent past. The entire region is covered in a layer of nitrogen ice, with glacier-like features at the edges suggesting that this ice slowly flows across the surface. In contrast, Cthulhu Regio is a dark, heavily cratered landscape, probably unchanged for billions of years. Its brownish-red colour is thought to be due to tarry chemicals called tholins, created as solar wind particles trigger reactions between methane and nitrogen in the atmosphere. Elsewhere, mountains (most likely made of frozen water ice) rise up to 4 km (2.5 miles) high, perhaps somehow pushed up from a hypothetical ocean lying just beneath the surface.

Charon

Pluto has a remarkably complex system of moons for an object of its size. The largest, Charon, is just over half the diameter of Pluto itself, and orbits in just 6.4 days at a distance of a mere 19,600 km (12,180 miles). As with many planetary satellites, tidal forces have slowed Charon's rotation so that it keeps the same face permanently turned towards Pluto. In this case, however, the tides have done a similar job to Pluto, so that one hemisphere is locked facing Charon.

Charon is thought to have formed in a similar way to Earth's moon, as debris thrown out from a collision with another large body coalesced in orbit around Pluto. The terrain is generally greyer than Pluto's, and appears to be dominated by frozen water rather than the more volatile ices seen on its larger neighbour, but there is a prominent reddish region around the north pole most likely covered in tholins. A mix of terrain types includes smooth plains, deep canyons and ice mountains, showing that Charon's past has been just as active as Pluto's.

Pluto's moons

Beyond the orbit of Charon, Pluto has four other known satellites – Styx, Nix, Kerberos and Hydra (in order of distance from the planet). All four are elongated along one axis, with Hydra and Nix, at 55 and 42 km (34 and 26 miles) long respectively, considerably larger than Kerberos and Styx at 12 and 7 km (7.5 and 4 miles) long.

The orbits of these moons are almost perfectly circular and aligned with Pluto's equator. This indicates that they are not captured bodies, but instead coalesced from a ring of debris following the same major impact that created Charon. Hydra and Kerberos both have two very distinct lobes, suggesting that they formed when a pair of smaller bodies collided, and the same is probably true of Styx and Nix. Orbit diameters range from 2.4 to 3.8 times larger than Charon's, forming a surprisingly compact system. The tidal forces exerted by Pluto and Charon are continuously changing as a result, causing the moons to tumble along their orbits with a chaotic rotation period.

New Horizons’ most detailed views of Pluto’s outer moons to scale with Charon

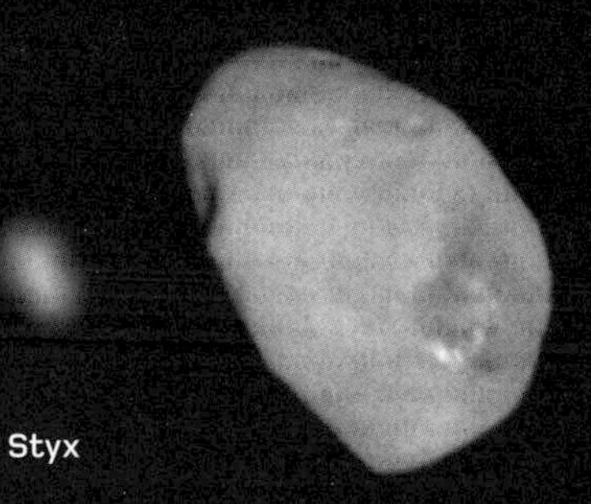

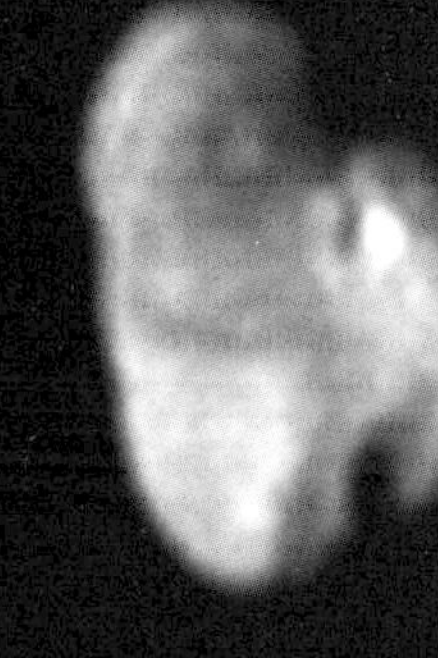

Albion

Although the existence of the Kuiper Belt was predicted as early as the 1940s (based on models of the solar system's formation), it remained purely theoretical until 1992, when the first new object beyond Neptune since Pluto was discovered. Known for a long time by the designation 1992 QB_1, it finally received an official name, 15760 Albion, in 2018.

With a diameter of around 140 km (87 miles), Albion is hard to study from Earth, but its orbit – and those of 2,400 other 'Trans-Neptunian Objects' now known – can clarify our picture of the outer solar system. Albion's 289-year path around the Sun is much less elliptical (ranging between 40.8 and 46.6 AU) and also less inclined than Pluto's, tilted at just two degrees from the plane of the solar system. Such orbits are common in the 'classical' Kuiper Belt, and rather different from that of Pluto itself. Hence, astronomers divide the KBO population into 'cold' orderly objects like Albion (sometimes referred to as 'cubewanos', from 'QB_1'), and 'hot', more eccentric ones like Pluto ('plutinos').

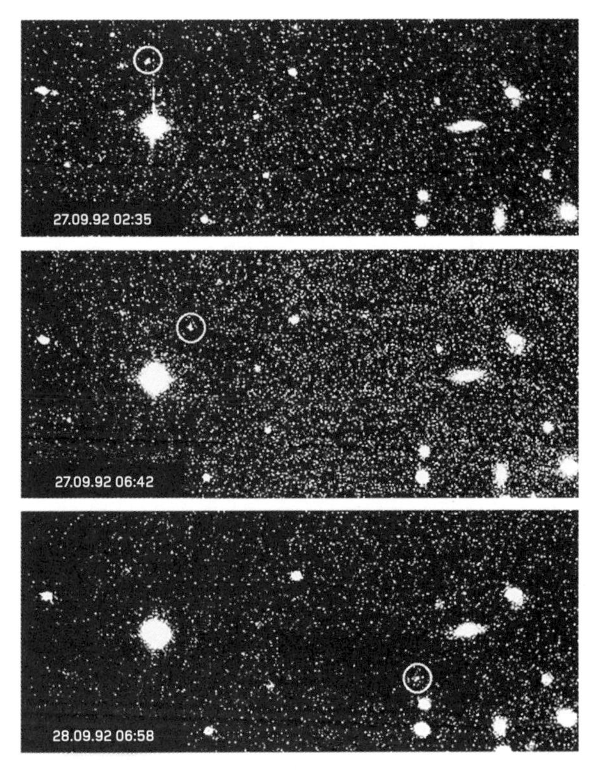

This sequence of images was used to discover Albion (ringed) in 1992.

Varuna and Ixion

The Kuiper Belt Objects 20000 Varuna and 28979 Ixion were two of the first more substantial ice dwarfs to be discovered, in 2000 and 2001 respectively. Varuna proved to be another cubewano, with a roughly circular orbit at around 43 AU from the Sun, while Ixion is a plutino – an object in an inclined, Pluto-like orbit, ranging between 30 and 49 AU.

Steady changes to Varuna's brightness indicate that it rotates once every 6.34 hours, with an elongated shape that reflects more or less sunlight depending on its orientation. Both worlds are red in colour, but there is a distinct difference in their surface brightnesses – Varuna is extremely dark and reflects back just 4 per cent of the light striking it, while Ixion is lighter, reflecting back 15 per cent. Spectroscopic analysis of Ixion's light has even revealed some of the materials on its surface. They include pure carbon, and also complex tholin chemicals, formed as solar wind particles alter carbon-based molecules in its surface ices.

An artist's impression of the elongated KBO Varuna

Quaoar

When it was found in 2002, 50000 Quaoar became the largest known KBO after Pluto. Images from the Hubble Space Telescope resolved the shape of its disc, allowing its diameter to be estimated at 1,260km (783 miles) – it is now recognized as a dwarf planet. Unlike Pluto, however, Quaoar is dark, with a surface that reflects just 9 per cent of light that strikes it. Its overall colour is reddish, although not as red as many smaller KBOs. In 2004, traces of water ice were detected on its surface, followed by methane and ethane ices.

Quaoar is another cubewano, with a roughly circular, low-inclination orbit ranging between 42.0 and 46.3 AU from the Sun. Cubewano orbits concentrate in this region to avoid orbital resonances with Neptune at around 39.5 and 48 AU. Objects that stray into the danger zones beyond find their orbits altered by repeated interaction with the giant planet; at the inner edge they may stabilize in more eccentric, Pluto-like orbits, while at the outer edge they can be ejected into the distant 'scattered disc'.

Haumea

Discovered in 2004, the Kuiper Belt Object 136108 Haumea has about one-third the mass of Pluto and is classed as a dwarf planet. In this case, however, the designation does not mean that Haumea is spherical – variations in its reflected light indicate that it is an elongated 'ellipsoid', with dimensions of around 2,100 × 1,600 × 1,100 km (1,300 × 1,000 × 700 miles), spinning on its axis once every 3.9 hours.

Haumea's shape and its speedy rotation are linked – its equator bulges outwards due to rapid motion that weakens the inward pull of gravity. The best explanation for the fast rotation itself is a violent collision in Haumea's past, which left the rocky core mostly intact while scattering fragments of the icy mantle to create a 'collisional family' of objects in related orbits. The collision is also thought to have created Haumea's two known satellites – Namaka and Hi'iaka. A destructive impact between other small moons probably generated the raw materials for the thin ring that was discovered around Haumea in 2017.

Hi'iaka
Haumea
Namaka

Eris

The large Trans-Neptunian Object 136199 Eris is the best-known member of the 'Scattered Disc' – a group of objects in highly elongated and inclined orbits with a separate evolutionary history from the plutinos and cubewanos of the 'classical Kuiper Belt'. First catalogued as 2003 UB_{313} (and nicknamed 'Xena'), Eris was named for the Greek goddess of discord following its designation as a dwarf planet and the 'demotion' of Pluto.

Eris was discovered near the outer edge of a long 558-year orbit that carries it between 38 and 98 AU from the Sun, but has, nevertheless, been subject to intense study. Eris is almost the same size as Pluto, but in contrast to most KBOs, its surface is distinctly grey and highly reflective, probably due to surface ice. The 2005 discovery of Dysnomia, a satellite in a 15.8-day orbit, revealed Eris's mass to be 27 per cent greater than Pluto's, suggesting that it contains much more rock than ice. This could generate sufficient heat to power geological activity and perhaps even warm a subterranean ocean.

An artist's impression of
Eris and Dysnomia illuminated
by distant sunlight

Makemake

The dwarf planet 136472 Makemake is in an oddity among large KBOs. The general shape of its 310-year orbit, taking it between 38.5 and 53.1 AU from the Sun, puts it in the 'classical Kuiper Belt', but its inclination of 29 degrees from the plane of the solar system means it is considered dynamically 'hot'. In other words, it did not form in its present orbit, but was instead scattered here by a past interaction with some other object.

At about three-quarters the size of Pluto, Makemake is one of the brightest KBOs. Variations in its brightness, due to surface markings, reveal that it spins every 7.7 hours. Nitrogen ice is not as widespread as on Pluto, but frozen methane is present in large grains, along with ethane and reddish tholin chemicals. Against expectations, Makemake seems to lack even a low-pressure atmosphere. A small moon, currently saddled with the ungainly moniker S/2015 (136472) 1, was announced in 2016 – evidence for a satellite had previously remained elusive because, in stark contrast to Makemake itself, this moon is as black as charcoal.

Makemake was expected to have a thin atmosphere similar to Pluto's, but the abrupt disappearance of stars passing behind it, with no prior dimming, reveals that it is effectively airless.

Sedna

Touted as the first member of a previously hypothetical 'Hills Cloud' of icy objects that venture far beyond the Kuiper Belt (see page 26), 90377 Sedna takes some 11,400 years to orbit the Sun. At the time of discovery in 2003, it was a mere 89.6 AU from the Sun (closer, in fact, than Eris – see page 362). But after passing perihelion in 2076 at a distance of 76 AU, it will begin its long retreat to aphelion at an estimated 936 AU.

Sedna is the reddest object in the solar system – redder even than Mars. Its crust is thought to be dominated by methane, methanol and nitrogen ices, with a thick coating of carbon-based tholin chemicals providing its red hue. In 2014, astronomers announced the discovery of another object in a similar (though smaller) orbit, lending weight to the idea that the Oort Cloud has an inward extension close to the plane of the solar system (see page 370). Objects following these orbits must have been pulled into them from closer to the Sun, perhaps during a close encounter with a passing star early in the solar system's history.

An artist's impression shows sunlight reflecting on the icy surface of Sedna at the very edge of the known solar system.

Planet Nine?

Does another large planet await discovery in the darkness of the outer solar system? The prospect of such a major discovery has tantalized astronomers since the early 20th century, and Pluto's discovery in 1930 was the serendipitous outcome of a deliberate search for one such 'Planet X'. However, as improvements to mathematical modelling and computing power refined our understanding of solar system dynamics, earlier concerns about planets not behaving as they should gradually faded away.

In recent years the quest has reignited, with the suggestion that the orbits of some of the solar system's most distant objects are aligned by an unseen influence. One possibility is that these distant worlds are affected by the gravitational pull of a 'Planet Nine' with ten times Earth's mass. Perhaps there is an ice giant ejected from closer to the Sun during the planetary reshuffle produced by the Nice model (see pages 42–45), or even a rogue planet captured from interstellar space.

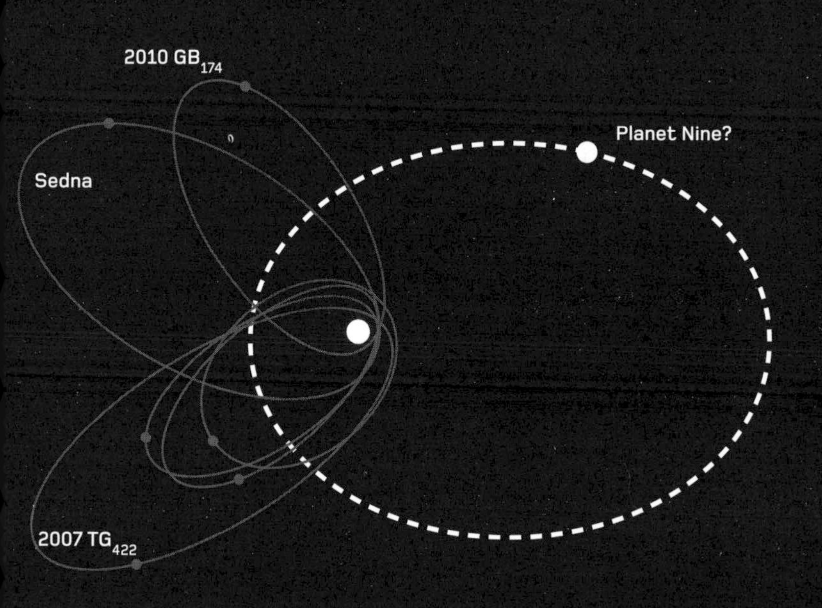

The orbits of the solar system's most remote bodies appear to cluster on one side of the Sun – are they influenced by the presence of an undiscovered ninth planet?

The Oort Cloud

The existence of a spherical cloud of comets surrounding the solar system was first proposed in order to explain the fact that long-period comets approach the Sun from all directions, and that 'fresh' icy comets continue to appear. Today, the so-called Oort Cloud remains a theoretical necessity that eludes direct observation, but its inferred structure has grown more complex. About 80 per cent of its comets are now thought to reside in an inner, doughnut-shaped extension called the Hills Cloud (stretching from about 2,000 to 20,000 AU from the Sun). This is the reservoir into which most comets were ejected during encounters with the giant planets in the early days of the solar system. From here, occasional collisions and close encounters eject comets into orbits with aphelia in the outer Oort Cloud, perhaps three times further from the Sun. Here, solar gravity is so weak that their orbits are easily circularized by the influence of other stars, before further encounters either strip them away into interstellar space or send them plunging back towards the inner solar system.

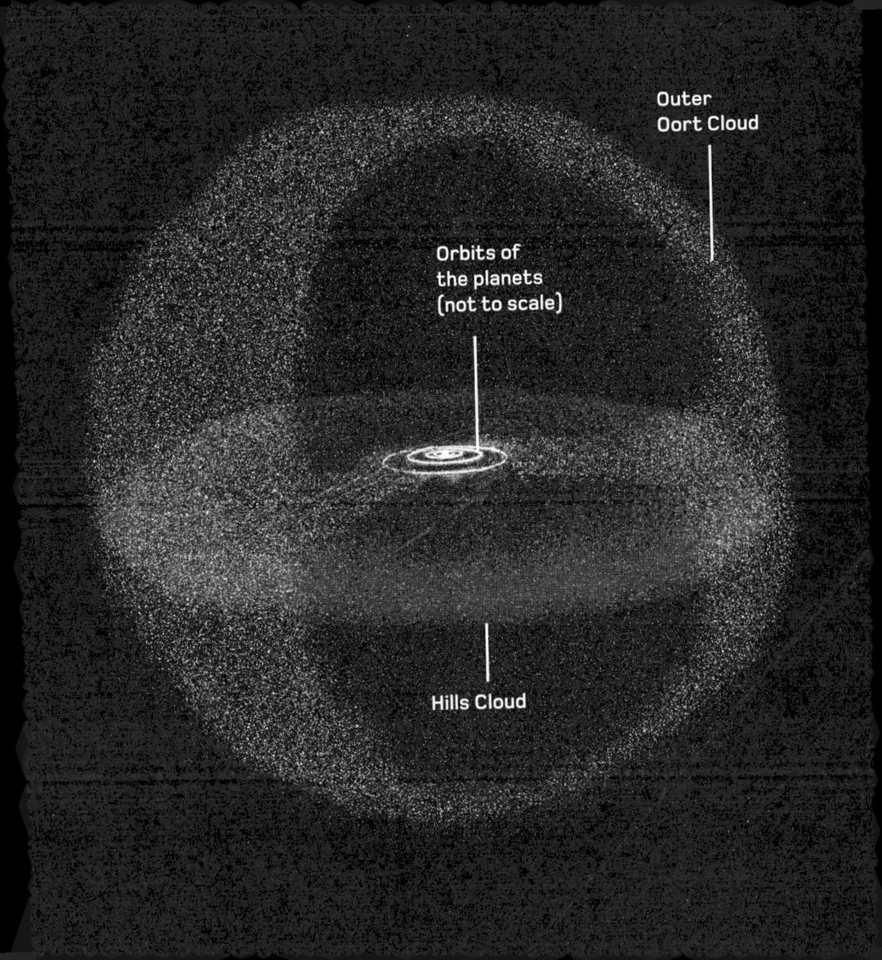

Outer
Oort Cloud

Orbits of
the planets
(not to scale)

Hills Cloud

Limits of the solar system

Most astronomers favour one of two definitions for the edge of the solar system. The more limited of these is the heliopause – the boundary at which the outward stream of the solar wind (see page 64) falters and dies. This boundary, encountered by the Voyager 1 space probe at 121 AU, defines the heliosphere within which the Sun's influence is paramount. Somewhat awkwardly, however, it suggests that many objects orbiting the Sun are actually *beyond* the edge of the solar system.

A more generous definition, then, is the Hill sphere, the region in which the Sun's gravity is capable of retaining bodies in orbit around it. An accurate calculation of the Hill sphere is complex – the influence of other stars (and even the vast distant mass of our galaxy's central hub) has to be taken into account. And even if an object moves into the sphere, there is no guarantee that it will *necessarily* be captured into orbit. Most estimates place the Hill sphere's outer edge about one light year (64,000 AU) from the Sun, but some put it at about twice that distance.

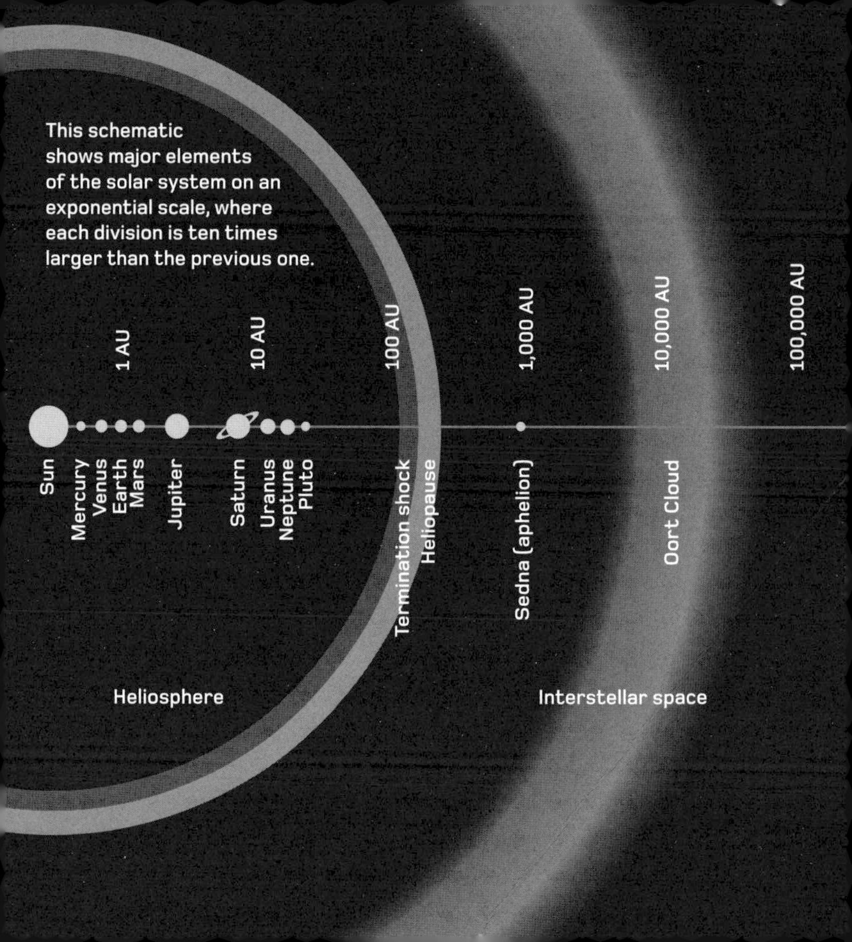

This schematic shows major elements of the solar system on an exponential scale, where each division is ten times larger than the previous one.

1 AU
10 AU
100 AU
1,000 AU
10,000 AU
100,000 AU

Sun
Mercury
Venus
Earth
Mars
Jupiter
Saturn
Uranus
Neptune
Pluto
Termination shock
Heliopause
Sedna (aphelion)
Oort Cloud

Heliosphere
Interstellar space

Exploring the solar system

For most of human history, astronomers have had to study the planets from afar. At first they simply tracked the planets' movements with the naked eye and using ingenious devices such as the astrolabe. Later they puzzled over the features they could see through telescopes. As the technology available for observing the sky improved, so too did the solar system's complexity, as new moons, asteroids and even entire planets were discovered.

The launch of the Sputnik 1 satellite in 1957 saw the beginning of a huge leap forward in our knowledge of the solar system. More powerful rockets and the electronic revolution soon meant that automated probes could venture beyond Earth's immediate orbit, visiting the Moon and our immediate planetary neighbours, before spreading out across the solar system. In the decades since, these scientific robots have developed many forms for exploring these alien worlds in our place, from orbiting surveillance platforms and atmospheric probes to sophisticated geological laboratories and intrepid rovers.

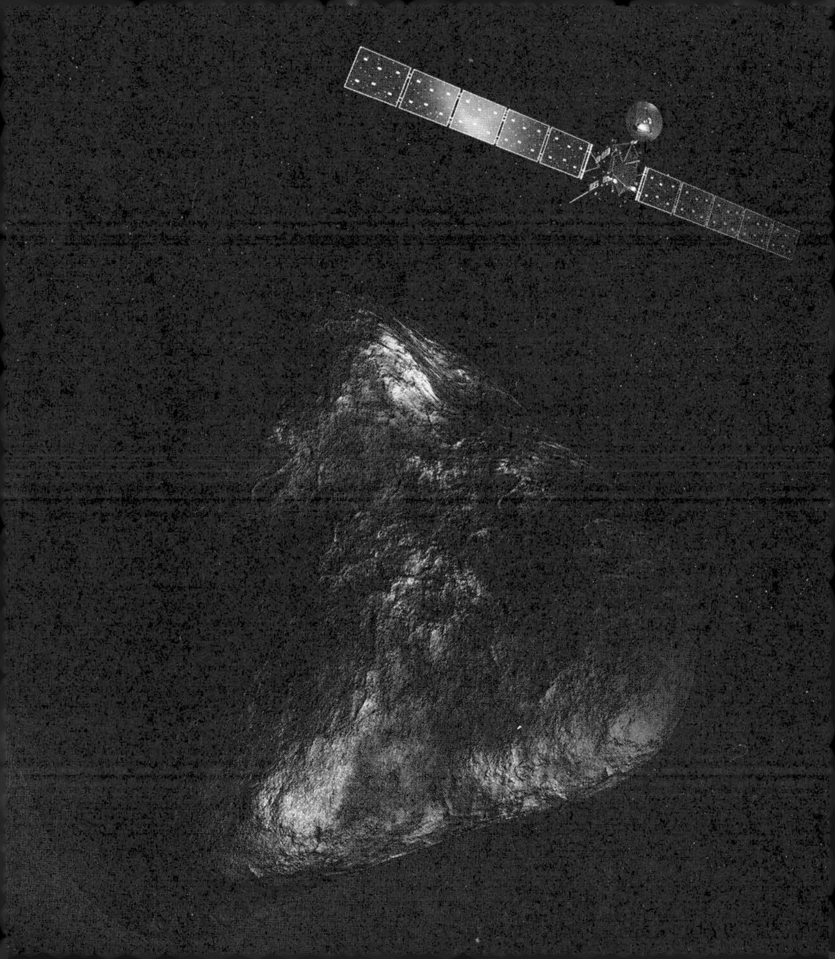

Lunar reconnaissance

In the late 1950s, the Soviet Union took an early lead in the 'space race' against the United States with a series of Luna probes that made flybys of the Moon or crashed into its surface. Putting a spacecraft in lunar orbit, however, was a far greater challenge since it required the use of braking rockets to slow the probe on arrival. Although this was first achieved in 1966 by the Luna 10 mission, the US space agency NASA soon had its own series of successful Lunar Orbiter probes mapping the Moon and looking for potential human landing sites. While the US crash-landing Ranger missions of 1964/65 sent back images that revealed the extent of lunar cratering for the first time, both sides struggled with attempts at a more difficult soft landing until February 1966, when Luna 9 sent back the first images from the Moon's surface. NASA followed Ranger with a series of Surveyor soft-landers that paved the way for the Apollo missions. Although beaten in the race to put humans on the Moon, the Soviets continued to launch increasingly sophisticated Luna missions into the 1970s, including two successful robotic Lunokhod rovers, and three automated 'sample return' missions.

The Apollo missions

Even after many decades, the Apollo programme remains an unsurpassed achievement in human spaceflight. The product of a concentrated national effort throughout the 1960s saw US astronauts Neil Armstrong and Buzz Aldrin land on the Moon on 20 July 1969. They reached the lunar surface using a sophisticated three-part vehicle launched on the Saturn V, still the most powerful rocket ever built. The Apollo 11 mission was followed by five more successful flights, in which ten more astronauts walked on the Moon between 1969 and 1972. While the driving force behind the Apollo programme was political, it produced a vast scientific bounty that transformed our understanding of the solar system as a whole. The six landing sites deliberately targeted interesting lunar geology, such as the Lunar Apennines, the boundaries between seas and highlands, and the ejecta blankets of major craters. Some 382 kg (842 lb) of carefully labelled rock samples were returned to Earth, allowing scientists to build up a precise timeline of lunar events, such as the Late Heavy Bombardment, which can be translated to the history of other worlds.

The Moon after Apollo

With the end of the Apollo programme, the focus of space exploration turned elsewhere and the Moon was largely neglected. Interest was reignited in 1994 by the US Clementine mission, a small satellite that returned some 1.8 million images of the lunar surface in just two months of operation. Clementine discovered subtle colour differences linked to certain minerals, and confirmed the existence of a huge depression, the South Pole–Aitken Basin, with permanently shadowed craters that might contain ice useful to future colonists. In 1998, NASA launched the Lunar Prospector probe into orbit. Its mission was to compile a comprehensive mineral map of the Moon, and to study lunar gravitational and magnetic fields. In the new millennium, other nations also took a renewed interest in the Moon. In 2003, the European Space Agency launched its own lunar probe, while 2007 saw the launch of lunar orbiters from Japan and China, with India following a year later. More recent NASA missions have focused on answering specific questions about the Moon's structure and particles in near-lunar space.

A 3D view of the lunar surface from Japan's Kaguya orbiter

Venusian landers and orbiters

After early failed attempts at a Venus flyby with the Venera 1 and 2 probes of the early 1960s, the Soviet Union set out to make the first landing on the surface. Veneras 3 through 6 were intended to parachute directly into the atmosphere upon arrival. Venera 3 lost contact during entry in March 1966 and failed to return any data, while Veneras 4, 5 and 6 returned data from the atmosphere, but lost contact before touching down. The more heavily armoured Veneras 7 and 8 (1970 and 1972, respectively) were the first to return data from the surface. Later missions took a different approach – Veneras 9 through 14 were delivered by an orbiting mothership that also acted as a radio relay, allowing images to be sent back for the first time. Answering fundamental questions about the Venusian landscape and geology, however, would require a different approach, using orbiters with radar-mapping equipment to pierce the clouds. The Pioneer Venus Orbiter and Veneras 15 and 16 made early contributions, but the most detailed maps of Venus originate from NASA's Magellan probe (1990–94).

Vikings on Mars

Following early investigations of Mars by Mariner probes, NASA launched a pair of ambitious missions to Mars in the mid-1970s. Each consisted of a two-part spacecraft – an orbiter to carry out a colour photographic survey of the red planet, and a lander that would send back pictures and environmental data from the surface, as well as collecting and analysing samples of the Martian soil. Vikings 1 and 2 arrived at Mars in summer 1976, with their landers touching down on the surface of Mars on 20 July and 3 September, respectively. Viking Lander 1 touched down on Chryse Planitia (see page 146) in what is now known to be a flood channel, while Lander 2 set down further north in a region known as Utopia Planitia, on a desert plain strewn with volcanic debris. Both landers monitored weather conditions, reported the composition of the atmosphere and analysed the surrounding terrain. Their most intriguing experiment searched for signs of microbial activity in the Martian soil – initial results from both landers seemed encouraging, but later attempts to repeat the tests met with negative results.

Pioneers to the outer solar system

While the inner solar system is relatively compact, the space beyond the asteroid belt is a different matter, with vast distances and long journey times between the giant planets. In 1964, however, NASA engineers realized that a rare planetary alignment in the late 1970s presented a unique opportunity for a 'Grand Tour' of all four outer planets, picking up speed at each through a technique called a 'gravitational slingshot'.

Planning began for the missions that would attempt the tour, but an understanding of conditions around Jupiter and Saturn, and a test of the slingshot principle, were key to its success. To this end, NASA launched the Pioneer 10 and 11 missions. In December 1973, Pioneer 10 became the first spacecraft to fly past Jupiter, revealing the planet's swirling cloudscapes in detail for the first time. Pioneer 11 followed a year later, executed its course change, and went on to return the first close-up pictures of Saturn, in September 1979. Both spacecraft are now on trajectories that will take them out of the solar system entirely.

DON
DAVIS

The Voyager missions

The spacecraft that eventually accomplished NASA's 'Grand Tour' of the outer solar system started life as modifications of the trusty Mariner template, after earlier and more ambitious schemes were scrapped in the early 1970s. They were ready for launch by mid-1977, with Voyager 2 launching first, and Voyager 1 setting off on a faster trajectory some 15 days later. By 1979, Voyager 1 had overtaken its sibling, and the two probes swung past Jupiter in March and July respectively, discovering the planet's rings, volcanic plumes rising from Io and the icy shell of Europa. Voyager 1's primary mission ended at Saturn in November 1980 – the trajectory required for its close Titan flyby could not also be used to fly on to Uranus. However, Voyager 2 continued onwards, executing a gravitational slingshot at Saturn, in August 1981. This put it on course to encounter Uranus in January 1986 and Neptune in August 1989, revealing them in detail for the first time. Both Voyagers are now leaving the solar system; in August 2012, Voyager 1 crossed the heliopause to became the first probe to send back signals from interstellar space.

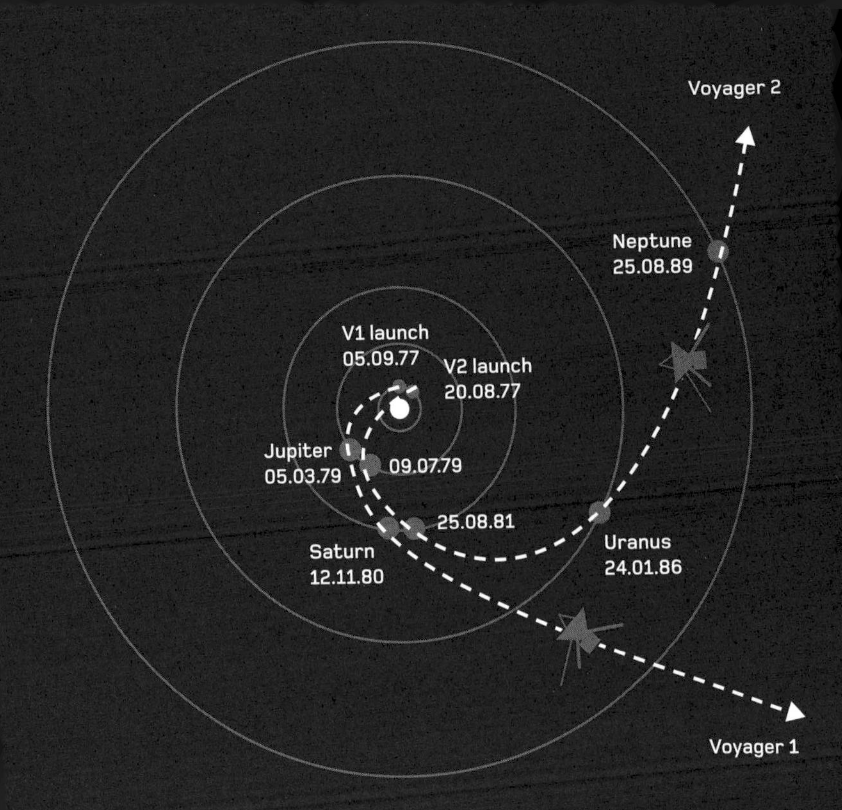

Trajectories of the Voyager probes

Missions to comets and asteroids

In the mid-1980s, an international armada of five probes was launched to greet the return of Comet Halley, perhaps the most famous of the solar system's smaller bodies. Two Japanese missions made long-range observations, while a pair of Russian craft flew closer to locate the nucleus. Finally, ESA's Giotto probe plunged into the coma, flying within 596 km (370 miles) of Halley's solid heart.

Since then, a wide range of missions have expanded our knowledge of both asteroids and comets. NASA's Galileo probe encountered main-belt asteroids en route to Jupiter in the early 1990s, while NASA's Stardust and Japan's Hayabusa returned samples of material from a comet and an asteroid to Earth laboratories (see pages 324 and 178). Longer-term studies include the NEAR-Shoemaker probe, which orbited asteroid 433 Eros for a year, and ESA's Rosetta mission, which accompanied comet 67P through its perihelion passage (see pages 190 and 326).

Return to Mars

Following the success of the Viking project, there was a long gap in the exploration of Mars – partly because space agencies' priorities were elsewhere and partly due to a string of mission failures that became known as the 'curse of Mars'. This began to change in 1997, when NASA successfully placed its Mars Global Surveyor (MGS) satellite in Martian orbit, and parachuted a lander called Pathfinder (equipped with a small robot rover called Sojourner) onto the surface. MGS delivered images at a far higher resolution than those delivered by Viking, revealing some of the first hints that liquid water might still be present on Mars. Subsequently, the pace of Mars exploration has accelerated, with successes such as the 2001 Mars Odyssey, which revealed vast amounts of ice beneath the northern plains, and Phoenix, which landed near the Martian north pole and monitored its descent into winter. Mars Reconnaissance Orbiter and ESA's Mars Express have continued MGS's work, photographing the planet in stunning detail, while the ExoMars Trace Gas Orbiter hopes to solve the mystery of Martian methane (see page 154).

Mars rovers

Some of the most important insights into Martian conditions past and present have come from NASA's robotic rovers. Following the success of Sojourner in 1997, a pair of much larger Mars Exploration Rovers – Spirit and Opportunity – landed on Mars in early 2004. Spirit's landing site was in Gusev, a crater whose floor was thought to be an ancient lake bed, but was subsequently found to be volcanic. In six years of exploration across 7.7 km (4.8 miles) of terrain, Spirit's discoveries included Martian dust devils and silica sand just beneath the surface, potentially formed by ancient hot springs. Opportunity landed in Meridiani Planum, an ancient shoreline, where it discovered rocks and minerals that probably formed under water. It has since travelled some 45 km (28 miles) during more than 15 years of operation. In August 2012, another even larger and more complex rover, Curiosity, touched down in Gale, an ancient crater filled with sediment deposits that were laid down underwater. Curiosity continues to investigate whether conditions in this region could once have supported life.

Galileo and Juno

The brief Voyager flybys of Jupiter provided our first detailed views of the giant planet and its complex moons, but they inevitably left many questions unanswered, and plans were being laid for a follow-up orbiter mission even before the Voyagers launched. Ultimately, this mission – Galileo – did not reach Jupiter until 1995 (taking a six-year route to the planet, in order to enter orbit around it). Galileo deployed a smaller probe into Jupiter's atmosphere before beginning a series of two-month loops around the planet that would bring it close to each of the major moons on several occasions. During eight years in orbit, it made a wealth of discoveries, such as the buried oceans on Ganymede and Callisto and the true extent of volcanism on Io. Its final act was a death plunge into Jupiter's atmosphere, in September 2003.

In July 2016, a new spacecraft arrived at Jupiter, aiming to reveal the secrets of the giant planet's atmosphere, magnetosphere, polar lights and internal structure. Juno occupies a highly elliptical, near-polar orbit that brings it to within 4,200 km (2,600 miles) of the cloud tops at close approach, providing unique views from hitherto unseen angles.

Cassini/Huygens

Following the Voyager flybys of Saturn, it was clear that the planet, its rings and varied system of moons held more than enough interest to justify a follow-up orbiter mission in the vein of Galileo. The bus-sized Cassini was to be the most ambitious space probe ever assembled, and became a collaboration between NASA and the ESA, with the latter primarily responsible for the Huygens lander that would parachute to the surface of Titan (see page 272).

Cassini arrived at Saturn in 2004, after a six-year journey involving multiple gravitational slingshots. It remained in orbit for some 13 years, making a series of discoveries, including at least eight new inner moons, intricate structures in Saturn's rings, the equatorial ridge on Iapetus and, most importantly of all, the enormous water plumes rising from Enceladus (see pages 246, 280 and 258). With its fuel supply running low, Cassini was deliberately deorbited into Saturn's atmosphere in September 2017, in order to protect the moons from possible contamination.

MESSENGER to Mercury

Entering orbit around Mercury presents a unique challenge owing to the planet's high orbital speed (see page 70). Mariner 10 achieved three initial flybys in the 1970s, using a slower-moving orbit that crossed over that of Mercury. However, matching orbit with the planet required NASA's MESSENGER (Mercury Surface, Space Environment, Geochemistry and Ranging) mission to follow a complex, seven-year flightpath involving one flyby of Earth, two of Venus and three of Mercury itself. By the time MESSENGER entered orbit in March 2011, it had already delivered a wealth of new data about the planet, including images of its previously unseen hemisphere.

MESSENGER remained in orbit until 2015, transforming our view of the solar system's smallest planet with evidence for surface water ice, carbon-based organic chemicals and a complex volcanic past (see page 78). BepiColombo, a joint European–Japanese mission, should continue to improve our understanding of Mercury when it arrives in the mid-2020s.

New Horizons

Launched in January 2006, NASA's New Horizons probe was already en route to its primary target of Pluto before the International Astronomical Union made its decision to reclassify the largest Kuiper Belt Object (KBO) as a mere dwarf planet. Nevertheless, the mission has been a stunning success, turning many previous ideas about KBOs on their heads.

Racing to catch Pluto on the inner edge of its orbit, and before its thin atmosphere had a chance to condense back into surface ice, New Horizons was the fastest spacecraft ever launched. It departed Earth at a speed of 16.26 km/s (36,373 mph) and boosted its speed still further with a gravitational slingshot at Jupiter, allowing it to reach Pluto in less than a decade. Over a few short days around its closest approach, it sent back images that revealed Pluto and its giant moon Charon as complex worlds full of unexpected geological activity (see pages 346–51). Following this encounter, it was set on course for a flyby of a much smaller KBO, designated (486958) 2014_{MU69}, in January 2019.

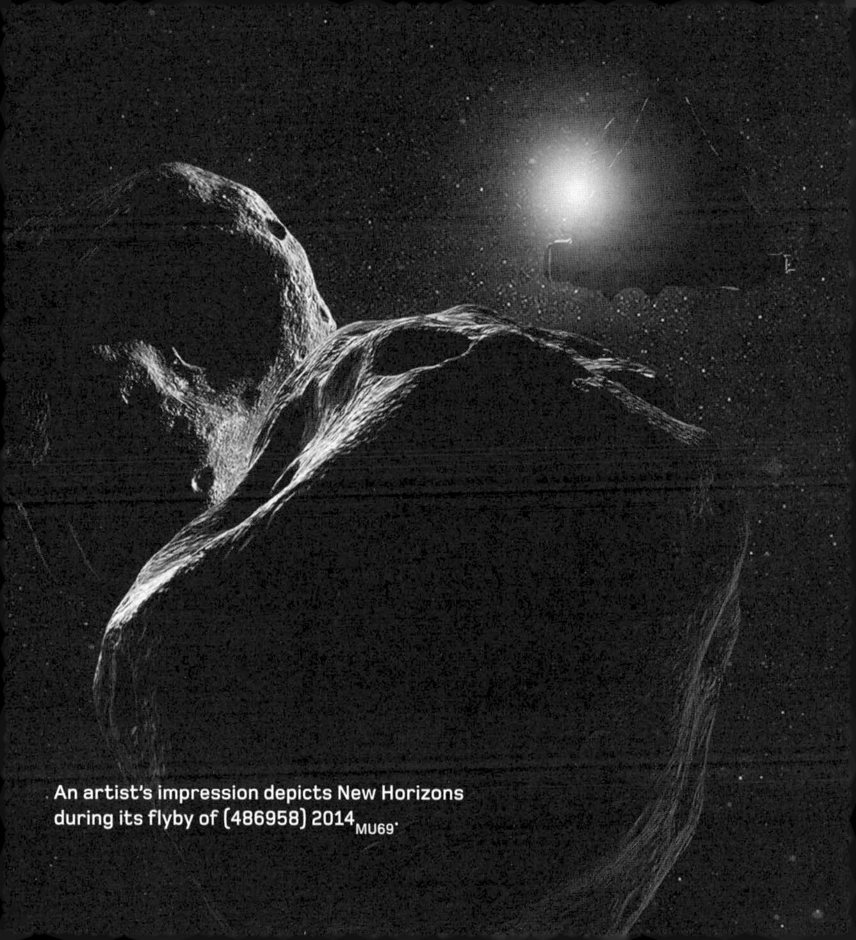

An artist's impression depicts New Horizons during its flyby of (486958) 2014 $_{MU69}$.

The future of manned exploration

Aside from the Apollo missions to the Moon some half a century ago, human exploration of our solar system has mostly been confined to Earth orbit. After decades of false starts, however, this is finally about to change. The growth of commercially operated spaceflight into Earth orbit has allowed the US space agency NASA to refocus on the exploration of other worlds. A new spacecraft, called Orion, will enter operation in the early 2020s – its advanced design can sustain a crew of four astronauts on missions of up to three weeks, and potentially much longer with additional modules attached. Orion's proposed missions include a return to the Moon, the establishment of a space station in lunar orbit, investigation of near-Earth asteroids and, ultimately, a manned expedition to Mars orbit in the 2030s. Meanwhile, the China National Space Administration is developing its own plans to send 'taikonauts' to the Moon and beyond, and commercial companies, such as SpaceX, are pioneering a new generation of giant rockets that, they claim, could put the first humans on Mars within a decade.

NASA's new spacecraft, the Orion Multi-Purpose Crew Vehicle, is designed with interplanetary exploration in mind.

Our future in the solar system

While many space advocates argue that colonization of the solar system is humanity's best hope of avoiding future extinction, the settlement of other worlds presents a huge challenge. Colonists would need to sustain themselves without a constant supply of resources from Earth, and water would play a key role; not just for drinking and growing crops, but also for processing to provide fuel and oxygen for breathing. Bases on our Moon might harvest ice from shaded craters at the poles – the airless environment would limit the scope of settlement, but a lunar base with lower gravity than Earth would make an ideal 'jumping off point' for exploration further afield. Mars offers more hospitable conditions, with plentiful ice in permafrost across much of the planet. In the long term, some argue, the planet could even be 'terraformed' to create a warmer, wetter world with a thicker atmosphere. Taming the hostile environment of Venus or settling on the moons of the outer planets would present even greater challenges, but if we wish to survive in the truly long term, humanity must go into the solar system.

This artist's impression depicts stages in the 'terraforming' of Mars to provide a habitable environment for human settlers.

Glossary

Asteroid
One of the countless rocky worlds of the inner solar system, mostly found in the main Asteroid Belt between the orbits of Mars and Jupiter.

Astronomical unit (AU)
A unit of astronomical measurement equivalent to Earth's average distance from the Sun – about 150 million km (93 million miles).

Centaur
A small icy world that orbits between the giant planets of the outer solar system.

Comet
A small chunk of rock and ice that typically orbits as a deep-frozen nucleus in the Oort Cloud or Kuiper Belt. When comets are disturbed, they can fall into elliptical orbits that bring them closer to the Sun, when their surfaces become active and produce streams of gas and dust.

Dwarf planet
A solar system object that, though large enough to otherwise qualify as a true planet, does not possess enough mass and gravity to clear its orbit of other bodies.

Eccentricity
A measure of how stretched or 'elliptical' an orbit is – perfectly circular orbits have an eccentricity of 0, while paths with an eccentricity of 1 or greater are open trajectories on which a body can escape orbit entirely.

Gas giant
A huge planet, far larger than Earth and dominated by a deep atmosphere of lightweight elements.

Gravity
A force of attraction that acts between all objects with mass.

Heliosphere
The region of space where the solar wind of particles blown out from the Sun is consistently streaming outwards.

Hills cloud
An doughnut-shaped inner extension of the Oort Cloud, between about 2,000 and 20,000 AU from the Sun.

Hill sphere
The region of space in which the Sun's gravity is the dominant force, stretching to at least 1 light year from the Sun.

Ice dwarf
An icy world, larger than a comet and including some dwarf planets, orbiting in or around the Kuiper Belt.

Ice giant
A planet larger than Earth whose interior is dominated by slushy 'ices' of chemicals such as water, ammonia and methane.

Inclination
The angle at which an orbit is tilted compared to the plane of the solar system or (in the case of satellites) the equator of the parent planet. Objects with inclinations greater than 90 degrees follow retrograde orbits.

Irregular satellite
A natural satellite in an eccentric, highly inclined or retrograde orbit that shows it did not form in orbit around a planet, but was captured later by its gravity.

Kuiper Belt
A doughnut-shaped ring of ice dwarfs and comets orbiting beyond Neptune.

Lagrangian point
One of several 'sweet spots' in a system involving two bodies with substantial gravity, where the influence of the object is neutralized and a third object can sustain a stable orbit around the more massive body.

Light year
A unit of astronomical measurement equal to the distance travelled by light in one year – roughly 9.5 million million km (5.9 trillion miles), or 64,000 AU.

Meteor
A short-lived 'shooting star' created when a small fragment of dust or ice enters a planet's atmosphere and burns up due to friction.

Meteorite
A fragment of rock from space, originating as part of an asteroid, that survives entry into a planet's atmosphere and makes it to the surface.

Moon
When capitalized, the name of Earth's single natural satellite. By extension, a moon (uncapitalized) is any natural satellite of a larger solar system object such as a planet, dwarf planet or asteroid.

Near-Earth Object
An asteroid or comet whose orbit lies close to Earth's, and which typically spends much of its time closer to the Sun than Mars.

Nebula
Any cloud of gas or dust floating in space. Nebulae are the material from which stars are born, and into which they are scattered again at the end of their lives.

Oort Cloud
A spherical shell of dormant comets, up to two light years across, that surrounds the entire solar system out to the limit of the Sun's Hill sphere.

Orbit
The typically elliptical path that a less massive body follows around a more massive one under the influence of gravity. Circular orbits are just an unusual form of ellipse.

Orbital period
The length of time taken for any astronomical object to complete a single orbit.

Planet
An object that follows its own orbit around the Sun, has enough mass and gravity to pull itself into a spherical shape, and which has cleared the region around it of other objects in long-term stable orbits (apart from its own satellites).

Planetary nebula
An expanding shell of gas thrown off by a Sunlike star at the end of its life.

Protoplanetary nebula
A dense, doughnut-shaped cloud of gas and dust left behind after the formation of a star, out of which planets and other objects coalesce in orbit.

Regular satellite
A natural satellite in a roughly circular orbit with low inclination to its planet's equator, which probably formed in orbit around the planet.

Resonance
A relationship between two objects orbiting a third, in which the orbital period of one object is a simple fraction of the other. As a result, the objects frequently return to the same alignment.

Retrograde orbit
An orbit in the opposite direction to the 'normal' patterns found in the solar system: for example, an asteroid or comet that orbits the Sun in the opposite direction to the planets, or a moon that orbits in the opposite direction to its planet's rotation.

Rocky planet
A relatively small planet predominantly composed of solid matter – sometimes known as a terrestrial planet due to their resemblance to Earth.

Satellite
Any object in orbit around a planet or smaller solar system objects. Satellites may be natural or artificial, and natural satellites may be regular or irregular.

Star
A huge ball of gas, compressed under its own gravity, that generates light and other forms of energy through nuclear reactions in its core.

Trojan
One of a group of asteroids that orbit the Sun at the same distance as Jupiter, and are found in clouds at two of the Lagrangian points of the Jupiter–Sun system.

Index

First published in Great Britain in 2018 by

Quercus Editions Ltd
Carmelite House
50 Victoria Embankment
London EC4Y 0DZ

An Hachette UK company

Text by Giles Sparrow

Design and editorial by Pikaia Imaging
Edited by Anna Southgate

A CIP catalogue record for this book is available from the British Library

PB ISBN 9781786485854
EBOOK ISBN 9781786485861

10 9 8 7 6 5 4 3 2 1

Printed and bound in China

Picture Credits: 9: NASA; 15: NASA/JPL-Caltech/ESO/R. Hurt; 19: NASA; 21: NASA/JPL-Caltech/SwRI/MSSS/Gerald Eichstädt; 23: NASA/JPL/Space Science Institute; 25: ESA/MPS/OSIRIS Team; 27: ESO/L. Calçada/Nick Risinger; 29: NASA, ESA, and The Hubble Heritage Team (STScI / AURA) ; 31: X-ray: NASA/CXC/Rutgers/J.Hughes; Optical: NASA/STScI ; 37: NASA; 39: NASA/JPL-Caltech; 45: ESO/S. Brunier; 47: NOAA; 49: NASA/JPL-Caltech/University of Arizona; 51: NASA/SDO (AIA); 53: Institute for Solar Physics. Observer: Tomas Hillberg. Image processing: Mats Löfdahl; 59: ESA ; 61: NASA; 65: NASA/ESA/SOHO; 69: NASA/Johns Hopkins University Applied Physics Laboratory/Carnegie Institution of Washington; 75: NASA; 77, 79: NASA/Johns Hopkins University Applied Physics Laboratory/Carnegie Institution of Washington; 81, 83, 85,87, 89, 91: NASA/JPL; 93, 95: NASA; 105: NASA/Robert Simmon; 111: NASA; 113: ESO/C. Malin; 115: Captmondo via Wikimedia; 117: USGS/D. Roddy; 119: USGS; 121: NASA/Goddard/Lunar Reconnaissance Orbiter; 123: NASA; 125: NASA; 127: NASA/JPL-Caltech; 129: Lunar and Planetary Institute, Lunar Orbiter Photo Gallery; 131: NASA; 133: Joe Huber via Wikimedia ; 135: NASA/GSFC/Arizona State Univ./Lunar Reconnaissance Orbiter; 137: NASA; 141: NASA, J. Bell (Cornell U.) and M. Wolff (SSI); ; 143: NASA/JPL/USGS; 145: ESA; 147: NASA/JPL/USGS; 149: ESA; 151: Pikaia Imaging; 153, 155, 157, 159: NASA/JPL-Caltech/Univ. of Arizona; 165: ESA - P.Carril ; 169: NASA/JPL-Caltech/UCLA/MPS/DLR/IDA; 171: NASA/JPL/MPS/DLR/IDA/Björn Jónsson; 173: NASA; 175: NASA/JPL; 177: NASA; 179: ISAS,JAXA; 181: NASA/JPL-Caltech; 183: Arecibo Observatory/NASA/NSF; 185: Kevin Heider @ LightBuckets via Wikimedia; 189: NASA/JPL; 191: NASA/JPL/JHUAPL; 195: Keck Observatory; 199: NASA/ESA/A. Simon (Goddard Space Flight Center); 203, 205: NASA/JPL-Caltech/SwRI/MSSS/Gerald Eichstadt/Sean Doran; 207: NASA/JPL; 209: NASA/Jet Propulsion Laboratory-Caltech/Southwest Research Institute; 211: Hubble Space Telescope Jupiter Imaging Team.; 213: NASA/JPL/University of Arizona; 215: NASA/JPL/USGS; 217: NASA/JPL-Caltech/SETI Institute; 219: NASA/JPL; 221: NASA/JPL-CalTech; 223: NASA/JPL; 225: NASA/JPL/Brown University; 227: NASA/JPL/ DLR(German Aerospace Center); 229: NASA/JPL/University of Arizona; 231: NASA/JPL/Cornell University; 235: NASA, ESA and E. Karkoschka (University of Arizona); 239: NASA/JPL-Caltech/SSI; 241: NASA/JPL-Caltech/Space Science Institute; 243: NASA/JPL-Caltech/SSI; 245: NASA/JPL-Caltech/Space Science Institute; 247, 249, 251, 253: NASA/JPL/Space Science Institute; 255: NASA/JPL-Caltech/Space Science Institute; 257: NASA/JPL/Space Science Institute; 259: NASA/JPL-Caltech/SSI/Kevin M. Gill via Wikimedia; 261, 263, 265, 267: NASA/JPL/Space Science Institute; 269: NASA/JPL/University of Arizona/University of Idaho; 271: NASA/JPL/Space Science Institute; 273: ESA/NASA/JPL/University of Arizona; 275: NASA/JPL-Caltech/ASI; 277, 279, 281, 283: NASA/JPL/Space Science Institute; 287: NASA/JPL-Caltech; 291: Erich Karkoschka (University of Arizona) and NASA/ESA; 293: NASA/JPL; 295: NASA/JPL-Caltech/Kevin M. Gill via Wikimedia; 297, 299, 301, 303, 307, 309,311, 313, 315, 317: NASA/JPL; 299: NASA/JPL; 323: James W. Young (TMO/JPL/NASA); 325: NASA/JPL-Caltech; 327: ESA, Rosetta, NAVCAM; processing by Giuseppe Conzo; 329: NASA/JPL-Caltech/M. Kelley (Univ. of Minnesota); 331: H.A. Weaver, T. E. Smith (Space Telescope Science Institute), and NASA; 333, 335: NASA/JPL/University of Maryland; 343: E. Kolmhofer, H. Raab; Johannes-Kepler-Observatory, Linz, Austria; 345: TRAPPIST/E. Jehin/ESO; 347, 349, 351, 353: NASA/Johns Hopkins University Applied Physics Laboratory/Southwest Research Institute; 355: European Southern Observatory; 357: Tomruen via Wikimedia; 359: NASA and G. Bacon (STScI); 361: NASA, ESA, and A. Feild (STScI); 363: NASA, ESA, and A. Schaller (for STScI); 365: ESO/L. Calçada/Nick Risinger; 367: NASA, ESA and Adolf Schaller; 375: European Space Agency; 377: OKB-1 via Wikimedia; 379: NASA; 381: Kaguya's images via Wikimedia; 385: NASA/Roel van der Hoorn via Wikimedia; 387: Donald Davies via Wikimedia; 391: ESA/MPS/OSIRIS Team/Kevin M. Gill via Wikimedia; 393: NASA/JPL-Caltech; 395: NASA/JPL-Caltech/Cornell; 397: NASA/JPL-Caltech/SwRI/MSSS/Gerald Eichstädt/Seán Doran; 399: NASA/JPL; 401: NASA/Johns Hopkins University Applied Physics Laboratory/Carnegie Institution of Washington; 403: NASA/Johns Hopkins University Applied Physics Laboratory/Southwest Research Institute/Steve Gribben; 405: NASA; 407: Daein Ballard via Wikimedia.
All other artworks: Tim Brown/Pikaia Imaging.